The Predictors

The Predictors

THOMAS A. BASS

An Owl Book
Henry Holt and Company
New York

Henry Holt and Company, LLC
Publishers since 1866
115 West 18th Street
New York, New York 10011

Henry Holt ® is a registered trademark
of Henry Holt and Company, LLC.

Published in Canada by Fitzhenry & Whiteside Ltd.,
195 Allstate Parkway, Markham, Ontario L3R 4T8.

Portions of the book originally appeared in
The New Yorker in a slightly different form.

"Don't Fence Me In," by
Cole Porter © 1944 (renewed) Warner Bros. Inc.
All rights reserved. Used by permission.
Warner Bros. Publications U.S. Inc., Miami, FL 33014.

Library of Congress Cataloging-in-Publication Data
Bass, Thomas A.
The predictors / Thomas A. Bass.
p.   cm.
Includes index.
ISBN 0-8050-5757-9
1. Investment advisors. 2. Stock price forecasting.
3. Economic forecasting. 4. Investment analysis.
I. Title.
HG4621.B37   1999   98-41820
332.6—dc21   CIP

Henry Holt books are available for special promotions and
premiums. For details contact: Director, Special Markets.

First published in hardcover in 1999 by
Henry Holt and Company

First Owl Books Edition 2000

Designed by Michelle McMillian

Printed in the United States of America
1   3   5   7   9   10   8   6   4   2

To the Predictors

Contents

[handwritten margin notes:]

gn picture d: floor & computer trading; derivatives XΔz.

venture c'd & equity structure d startups used o val d lead ptnrs.

econ chaos Jul & Fin.

pp.79-82: chartism

pp 88-96: Farmer Pely do in si kony & ef mkt Jul.

p.106-7: pairs trading & stat arb

p.111 Buffett
p.112 Prechter

(handwritten annotations)

pp. 119-20: moral g? Predicam Co

117-8 & 121-7 Black-Schol model & ti by O'Connor Asso

160: stock index arb
161-3: econ f & 2 vary & 3 vary mkt

177-82: deNoyo

189-90: Soros ak £; O'Har strategy

190-1

202-9: problem d non-stationarity.

237-40 moral g? & structure d l firm

263-8 Derivativ & fin collapses: Leeson, P&G, Orange County, Soros

272-3: mkt friction i trading on futures

304-6 turn In l black box sve?

The Predictors

Turbulence

I want to ride to the ridge where the West commences,
Gaze at the moon till I lose my senses,
Can't look at hobbles and I can't stand fences,
Don't fence me in.

—COLE PORTER

All hell has broken loose in the Chicago exchanges. A speculative tsunami from somewhere in Asia has washed over the gold market in Zurich before crashing into the pits at the Mercantile Exchange. Hundreds of men and a handful of women, packed hip to butt in raked arenas, are yelling at one another and waving upraised hands, palms out, in a desperate bid to dump Eurodollars. The market is gapping downward faster than orders can get filled. The exchange computers have exploded. The three-story light board surrounding the pits is no longer tracking the action. The traders in bright red and yellow jackets look like life-vested sailors shouting to be saved from a churning market that is about to swamp them.

This is not a number day, when Commerce takes the temperature of the economy. This is not a triple witching hour when quarterly contracts expire. Neither the scalpers betting their own money nor the floor brokers trading corporate accounts had guessed that a tidal wave of sell orders would build all night through Asia and Europe before flooding into Chicago. Clinging to their bleachers, with a sea of trading cards rising at their feet, the Merc's red-faced traders are scrambling to get clear of a foreign exchange market so big that by the end of the day more than a trillion dollars will have changed hands.

To the uproar of shouting traders is added the sound of sirens going off every fifteen minutes and the booming of loudspeakers calling out incomprehensible messages. Surrounding the pits is a huddle of clerks running orders back to the thousand trading stations that overlook the floor. Here wire clerks, outfitted with multiple telephones draped around their necks, are doing their own agitated dance through the day's incoming orders. Surveying the scene from metal parapets suspended over the pits are the blue-coated officials who keyboard every wink and nod into an electronic ticker tape that flashes the numbers up on the wall and out from there to the four hundred thousand computer screens on which the world's financial traders are following the action on this eighth day of April 1996.

The price of admission for standing room in a Merc pit is half a million dollars. There are many ways to play the game, but really only two kinds of players. Speculators and hedgers. Those scalping profits from churning markets, and those seeking shelter from the storm. Ninety-seven percent of the daily churn in the financial markets is generated by speculators. This is the official figure, published by the Merc, which defends speculators as necessary for keeping the markets "deep and liquid." Speculators stir the pits and hedgers pump them with funds, and between the two of them one gets the feeding frenzy known as the world financial markets.

The open outcry system, where every broker is his or her own auctioneer and every price is discovered by yelling it to the world, is turning today into a mad howl of sellers searching for buyers. The sell virus has jumped from foreign exchange to interest rates and out from there to all the commodities that will cost more when the price of money goes up. Everyone is shouting and gesticulating. Palm out, *sell!* Palm in, *buy!* Fingers near the face indicate quantity. Fingers to the side of the body indicate price. Upright fingers count one through five. Horizontal fingers count six through nine. Single digits are counted on the chin. Tens are counted on the forehead. A "fill or kill" order looks like someone shooting himself in the temple. "Sell two thousand" is the horns of a cuckold.

Traders keep flashing check marks above their heads to verify the price at which their orders are being filled, and everywhere brokers are tugging their earlobes—the sign to fill an order no matter what the price. They are desperate to get a broker across the pit to give them a salute from the nose, meaning their order has been executed. All morning the shouts go out for ten lots, thousand lots, a *lot* being a futures contract that can cost up to 5 million dollars. Scalpers flap their hands, trying to attract butterfly spreads, or put their hands to their throats, searching for strangles. They keep patting themselves on top of the head, asking, "What size is the market?" They pump their elbows to signal market orders, limit orders, and canceled limit orders, which look like someone cutting his throat.

Big men in steel-toed shoes, who will lose their voices and retreat to the back office before they are forty, are jammed together so closely that they can tell what their neighbors ate for breakfast. They pull their jackets over their heads when someone farts. They goose one another after a good trade. They hurl catcalls at women being trained as runners. This is a hard-drinking, high-cholesterol crowd, barely graduated from toga parties, who today are flushing Euro-dollars down the drain and betting hard against the United States Treasury. The joke in Chicago is that these people, if they weren't running the world financial markets, would be driving taxis.

Two thousand miles away, down by the railroad yard in Santa Fe, New Mexico, sits an old Coca-Cola warehouse. Occupying the ground floor are a used-furniture dealer and a shop selling crystal balls and spiritual advice. The second floor, refurbished with sand-colored stucco walls and track lighting, holds a kitchen and bathroom, five former bedrooms converted to offices, and a central, sunlit space, from which rises a staircase leading to a rooftop gazebo. The big room is filled with hanging plants, a couch, a table, and a couple of computer terminals. The screens are glowing with the green sawtooth lines of financial markets ticking up, down, up, down.

Climbing the warehouse stairs, after eating lunch across the street at the Zia Diner, are a dozen men and women in sandals and shorts and T-shirts decorated with bespectacled banana slugs and Leonardo da Vinci's design for the first bicycle. Except for the stout Dave DeMers, who is bringing up the rear like Falstaff, and the wiry William, who is nut-brown from spending so many spring days in his kayak, the men are tall and lean. Equally fit are the two women among them, who are also dressed in shorts and sandals. Clara, the Bernese mountain dog who has one brown and one blue eye, bounds upstairs, swishes her tail, and gives a hearty *woof!*

"One woof means we're winning!" Doyne mugs, giving the dog a rub behind the ear. "Let's go see how you're doing."

He opens the door and crosses the room. The blinking computers tell him at a glance that Clara was right. All morning the machines have been selling short, dumping shares in markets predicted to fall.

"Not bad," says Norman, smiling at the day's winnings, which are tallied on the computer screen in front of them. "That was a million-dollar lunch."

"This calls for some celebrating," announces DeMers, who riffles through the soda cans and leftover birthday cake in the refrigerator to produce a bottle of champagne.

Beep! sings the computer, making its final trade for the day.

Later in the afternoon, they grab the champagne and head upstairs to their rooftop lookout. Below them lie the red adobe houses of Santa Fe, over which hangs a leafy gauze of cottonwoods and willows. Off to the east rise the snow-covered peaks of the Sangre de Cristo—Blood of Christ—mountains. Falling away to the west are the layer-cake mesas of the Rio Grande valley. The air is tangy with the smell of burning piñon. The setting sun is raking the desert into a fierce display of umbers and ochres, as a crescent moon rises in a cloudless sky. It is a good day to be alive. A good day to have an extra million dollars in your pocket.

For big players in the world financial markets, who often burn the midnight oil, they look pleasantly relaxed. No one is going to follow the Asian markets over the weekend or worry about Monday's opening prices. While their automated system hums along on its own, Sonia is going to climb Santa Fe Baldy; William is going to kayak the

Rio Grande Gorge; Doyne is going to ride his new mountain bike up Glorieta Mesa; Norman and his family will hunt for mushrooms behind Los Alamos; and Karen will plant her garden.

"Who knows, maybe we've finally cracked the problem," says Doyne, turning to Norman, who is stretched out beside him on a lounge chair. "In any case, it sure beats losing," he says with a grin on his face.

Norman raises his glass. "To our next million," he says.

Black Box

It is generally agreed that casinos should, in the public
interest, be inaccessible and expensive. And perhaps the
same is true of Stock Exchanges.
— JOHN MAYNARD KEYNES

Money is better than poverty, if only for financial reasons.
— WOODY ALLEN

The Chicago Mercantile Exchange, founded a hundred years ago as
the Butter and Egg Board, used to be famous for trading pork bellies
and cattle. Then, in the early 1970s, it expanded the idea of a commodity to include Deutsche marks, dollars, and other foreign currencies. Next came trading in stock market indexes, Treasury bills,
Eurodollars, cross-rates, straddles, rolling spot, and a host of other
bets designed to slice the time value of money into ever thinner
tranches. Bets invented by Nick the Greek and games formerly
played only in Las Vegas now employ four thousand hyperactive
traders at the Merc and many thousands more on the world's two
hundred other financial exchanges.

The Merc runs four major games in circular pits bulging with brokers stacked fifty deep. They are betting on commodities, stock market indexes, Eurodollars, or currencies. On the lower trading floor,
next to the old-timers barking over broiler chickens and hogs, is the
pit for trading indexes. Here people bet not on individual stocks but
on groups of stock meant to reflect the movement of entire markets.

The bet is further abstracted by turning it into a futures contract or an option to buy a futures contract three months from today. This bet-on-a-bet is known as a derivative, since the value of a futures contract is derived from the value of the underlying stock. Because the action is faster and the margins thinner—five percent down will buy you a futures contract on the DAX 30 in Frankfurt, the CAC 40 in Paris, the FTSE 100 in London, the Nikkei 225 in Tokyo, or the Standard & Poor's 500 in New York—trading in derivatives now swamps the markets on which they depend.

Upstairs at the Merc, which looks like a football field with dozens of teams and thousands of players fielded simultaneously, lies the quadrant reserved for trading foreign currencies. These pits share the floor with a wild arena devoted to betting on Eurodollars. The United States, thanks to its negative trade balance, hemorrhages greenbacks. This money floats offshore in a huge pool of extra-national capital known as the Euromarkets. Changes in domestic interest rates whipsaw the value of Eurodollars, and it is a fine game to hedge the two against each other. A banner hanging over the pit trumpets the fact that Eurodollars are now the world's most actively traded futures contract. Every upraised finger in this pit represents a million dollars, and this one pit alone trades seventy-five trillion dollars' worth of contracts a year.

Futures and options on futures are the chips in play at the Merc. People are placing a bet, or reserving the right to place a bet, that three, six, nine months from now a freight car full of pork bellies or the Financial Times Stock Exchange (FTSE, pronounced "Footsie") market index in London will rise or fall in price. What began as a market for agricultural products—a kind of farmer's insurance policy— has expanded over the years into a mirror world of speculators speculating on one another's speculations.

Rushing among the deck holders bristling with trading cards and the arbitrage clerks hustling orders from pit to pit are other traders who wear bright green jackets with black patches on their backs. These are the out-trade clerks—the head referees—who are converging on the yen pit to settle a dispute. In the Osaka and Tokyo exchanges, these referees, called *saitori*, blow whistles and call time-out.

Here in Chicago they are going to slap an unruly trader with a thirty-thousand-dollar fine.

After opening with a flurry of sell orders at 7:20 A.M. and spiraling downward all day, the markets are off one percent, a move they normally make in a month. Will prices bounce back? Will they fall through the floor? The action, already frantic, cranks itself up another notch before the closing bell strikes at 2:00 P.M. Some of these screaming traders will soon be back to driving taxis. Others are contemplating a bet known as the O'Hare straddle. You max out your credit line, make a stab at guessing tomorrow's opening prices, and flee to O'Hare Airport. You wake up tomorrow in Rio, either a bankrupt fugitive or a lucky millionaire.

The furious finger show at the Merc is also playing down the street at the Chicago Board of Trade. The CBOT ("See-Bot"), Chicago's oldest and second-biggest financial exchange, occupies the forty-five-story art deco building that anchors the southern end of LaSalle, Chicago's equivalent to Wall Street. The Chicago Board of Trade was founded in 1848 by grain merchants, which explains why a thirty-foot statue of Ceres, the Roman goddess of grain, graces the roof. But long ago the big game at the CBOT passed from grain to money, in this case, futures and options on U.S. Treasury bonds.

The hot action is in the thirty-year T-bond pit, an octagonal grandstand cobbled together from rough-cut plywood, where five hundred traders, packed shoulder to shoulder in steeply raked tiers, are betting on the cost of American debt thirty years from now. This has been a roaring good business since the Vietnam War, which was financed by debt, and which marked the turning point when America flipped from being the world's largest creditor to its biggest debtor. On a busy day like today, more than a hundred billion dollars in contracts will be swapped by traders wearing digital headdresses. The floor is a swirl of red, green, and yellow jackets that on closer inspection reveal themselves to be little more than polyester sacks slipped on for the purpose of identifying the person yelling in your face.

The most successful traders, many of them bull-necked men who played ball in college, stand in the back row of the pit, called the candy store. From here they can get a clear view of the prices being signaled below, the tote boards flashing along the walls, and the arbitrage clerks who are fielding incoming quotes from the wire houses whose desks surround the floor. The wire clerk at Merrill Lynch is scratching his nose, and dozens of locals start scrambling to get out in front of what they suspect will be a big trade. A bald guy is going berserk over an order he thought was filled, until his counterpart keeled over from heat stroke. Paramedics at the ready, the downed trader is hoisted out of the pit hand over hand like a plank of lumber.

The crowd lurches against the railing. A broker gets shoved off the candy store. His trading cards flutter to the floor, while he makes a wild grab for his colleagues' trousers. He heaves himself back into the scrum. Chopping hard with his elbows, he begins yelling for a bid on a thousand lots. He lays two fingers alongside his nose. "Goldberg! Goldberg!" he shouts to a trader from Goldman Sachs. "What do you bid?" Then from behind his head sprouts a full coxcomb of fingers, each one representing a thousand contracts.

The trader's clerk flashes the numbers back to their order desk, where a wire clerk with multiple telephones craned to his face is calling out a constant patter of prices to the big customers who want to know which way the market is moving. Another clerk is phoning the back office to run the theoreticals on flex options, overnight straddles, and other arcane bets that will hedge their position. Sitting on a raised platform over the pit, a CBOT market recorder keyboards the trade into the exchange's computers. Up on the light boards and out over the quote machines flashes the new price. In the flick of a decimal place, traders around the world begin scurrying for cover.

Without missing a beat, the big trader in the candy store goes looking for another sale. He holds up his hands and begins screwing a finger into his upraised palm. He is building a gallows. He is asking for a lynching, which, in this case, means a quote from Merrill Lynch. The traders next to him are using their own hand signals to solicit bids from Morgan Stanley (a big "M"), Bear Stearns (a bear hug), and Refco (smoking a reefer). The traders standing in the futures pit keep glancing into the neighboring options pit. They are tracking related markets that move together. They are looking for arbitrage opportunities, quick profits that can be scalped by buying contracts in one market and selling them in another.

The action at the CBOT is divided among pits trading futures contracts and options on T-bonds and Treasury notes, thirty-day Fed funds, and two-, five-, and ten-year Treasury notes. But there is really only one big game being played in the CBOT finance room. The bets are all on U.S. government debt, packaged in different ways, but dependent on the world's tolerance for allowing America to live beyond its means. Sixty-five billion dollars a day in U.S. government debt is traded in the Chicago futures markets. This is nearly double the average volume of trading on the New York Stock Exchange, which is the world's largest stock market. The magnitude and power of the bond markets apparently caught Bill Clinton by surprise when he was first elected President of the United States. "I used to think if there was reincarnation, I wanted to come back as the President or the Pope," quipped James Carville, Clinton's campaign manager. "But now I want to be the bond market. You can intimidate everybody."

Next to the CBOT finance room is another floor devoted to trading futures contracts and options on soybeans, corn, wheat, gold, and silver—the staples on which the Chicago Board of Trade was built. This is a gentleman's game, played in a big, well-lit room between 9:30 A.M. and 1:15 P.M. Today many of the commodities brokers are wearing silver neckties and purple shirts to celebrate a Northwestern victory over Michigan. When not standing with their fingers in the air, each one representing five thousand bushels of grain, the traders are reaching into their wallets to pay off sports bets or get a pool going on the next big game.

The traders are agitated over the rising prices. They send up a flurry of shouts, but the action is bigger and faster across Van Buren Street in another market, the Chicago Board Options Exchange, or CBOE ("See-Bo"), which is devoted to trading futures and options on the stock market. The idea of transforming stocks and bonds into commodities and swapping them like futures contracts on soybeans was the Big Bang that twenty-five years ago reinvented Chicago's markets. The transformation dates from 1971, when the world went off the gold standard. For the first time in history, every major country had a fiat currency whose value is based not on gold, silver, rice, or some other commodity, but solely on the information that flows through the currency as it floats on the world financial markets. Some of this information comes in the form of numbers measuring economic well-being or malaise. The rest comes from fear, greed, and the rumor mill known as *market sentiment*.

"In 1971 we switched from the gold standard to the information standard," declares former Citicorp chairman Walter Wriston, and nowhere were the satellite feeds, uplinks, transponders, and telephone lines required for trading on this information more densely packed than in Chicago. Every commodity has its market. Cotton is traded at the New York Cotton Exchange, oil at the New York Mercantile Exchange, soybeans in São Paulo, rubber in Singapore, silk in Kobe. But Chicago's three biggest exchanges—the Merc (one hundred and sixty-eight trillion dollars per year), the CBOT (twenty-six trillion dollars), and the CBOE (five and a half trillion dollars)—are those that mastered the art of trading money. The notional value of contracts traded on the New York Stock Exchange, by comparison, is about seven trillion dollars a year.

Borrowing the idea from Las Vegas, where the game was first played, the Chicago Board of Trade opened a pit in the early 1970s for betting on the price of individual stocks, like IBM and Texaco. Then it opened another pit for betting on the aggregate value of America's five hundred leading stocks. These games were run out of an old lounge called the Smoker, until city marshals threatened to close it down as a fire hazard. This spurred the exchange into taking its five-story trading floor and chopping it in half horizontally. The bond markets stayed downstairs; upstairs, on the new, thirty-thousand-square-foot trading floor, went the Chicago Board Options Exchange. By 1985 the CBOE was big enough for its own building. It moved across the street into a high-tech palace connected to the

CBOT by a pedestrian bridge that doubles as an electrical umbilical cord.

It has two big pits devoted to trading stock market indexes, but most of the CBOE floor is packed with what look like jungle gyms dripping with video monitors. The screens flash the prices of the five hundred individual stocks whose futures and options are traded at the CBOE. Brokers cluster at these trading posts, eyeballing the numbers, searching for the ticking decimals that will launch them into shouting out prices and sales.

The action is a fast-ball series of round-robin plays in which brokers buy options on stocks and sell futures contracts on the stock index. An opportunity to buy at price y is banked against a commitment to sell at price x. The name of the game is statistical arbitrage. Stat arbs are looking for "mispricings," little bumps in the road that are produced when one market lags behind or edges ahead of another. When they see a bump, the stat arbs rush in to pick up a nickel here, a dime there, until the well-oiled world markets get back to rolling in synch.

Over in the CBOE index pits a bald trader in a lime green jacket is yelling to buy "two thousand SPX strangles with a PM settle."

Next to the bald trader, someone wearing a Hawaiian floral-print jacket begins shouting for "one-hundred-and-five-percent-out-of-the-money calls."

"No, I don't want the quote in dollars," he signals emphatically. "I want it in percentage of spot."

All day this talk of strangles and spots is carried on through people's hands. All day the markets move, for you, against you, but seldom for any reason related to the news that races in lights across the wall. *The Wall Street Journal* will report the markets moved because people thought interest rates were going up or down or because the President sneezed. But traders in Chicago know better. They know the markets move because the markets move. They are agitated by political events and rumors. They are ruffled by fads, but under these oscillations run deeper currents, like ocean tides, that flow through the world markets in great upwellings and cycles that have unforeseeable and unstoppable rhythms of their own.

Rising above Chicago's big financial exchanges are skyscrapers filled with gray-carpeted corridors guarded by surveillance cameras, motion detectors, and code boxes. Off these corridors lie the trading floors of companies active in the markets below. Many of these offices are nothing more than rabbit warrens of desks piled with banks of computer terminals and video screens. The desks are littered with coffee cups, spreadsheets, and half-eaten plates of food delivered to traders who are too busy to move. People are shouting into telephones and squawk boxes. They are getting juiced on the markets, buzzed by the sheer terror and delight of trading a trillion dollars a day in foreign exchange or pumping up a derivatives market whose nominal value in 1998 was seventy trillion dollars. This is eight times larger than the annual gross national product of the United States of America.

Among the corporate trading floors rising over LaSalle Street, the biggest and by far the most impressive belongs to Swiss Bank Corporation. SBC is a venerable institution, based in Basel, whose quaint logo of three crossed keys belies the fact that it is one of the most aggressive players in the world financial markets. It has proved so successful at trading derivatives and other speculative instruments that SBC, Switzerland's third largest bank, was able to merge as the dominant partner with Union Bank of Switzerland, the country's largest bank, in 1998. The merger created a combined entity, known as UBS, whose seven hundred billion dollars in assets and one trillion dollars under management made it the world's largest bank.

The Swiss Bank trading room is none other than the top two floors of the old Chicago Board of Trade, which was redesigned to house the Chicago Board Options Exchange. In 1988, after the CBOE moved across the street, the room became the trading floor for O'Connor & Associates, Chicago's most successful derivatives dealer. Then in 1992, after O'Connor allowed itself to be bought by Swiss Bank, the room got redecorated with Swiss flags.

SBC does a lot of business downstairs on the Chicago exchanges. It does even more business on the New York and London exchanges. But it does most of its business in the interbank market. This is the off-exchange, multitrillion-dollar betting pool organized by the world's dozen major banks. By wiring themselves together into what was effectively the world's first global computer, the banks have eliminated the middlemen. They no longer need the thousands of exchange-based traders whose yelling gets in the way of playing the numbers straight up. Forex, derivatives, futures, options, and even your basic stocks and bonds are all in the process of moving away from exchange-based trading to internets.

Money today is bits and bytes that sweep around the world through satellite transponders and fiber optic cables. Money is information that travels at the speed of light, taking a hundred milliseconds to get from Tokyo to Chicago. When money increases in speed, it also increases in volume. The satellites that beam money information between continents produce a bulge of credit that rotates with the sun around the planet. By moving trillions of dollars that used to remain stationary, satellites have produced an estimated five percent jump in the world credit supply.

When the markets close at the end of the day in Chicago, the book is passed to Osaka and Hong Kong. Then when the Pacific rim signs off for the night, the action moves to the LIFFE (London International Financial Futures Exchange) or SOFFEX (Swiss Options and Financial Futures Exchange). These markets all speak the same language and operate under the same rules, beginning with the Law of One Price. The slightest edge—a cocoa bean selling for an eighth of a cent less in Rio than in New York, or a Japanese warrant down a tick in London—will be arbitraged to parity in seconds. The next step in the optic integration of global finance will be to skip these exchanges altogether, which will happen when the globe is wired into one big casino played nonstop by a worldwide web of computers.

To play the interbank game one needs an AAA credit rating and deep pockets. One can bet the bank and lose it, like Barings and Orange County, or leverage oneself into a bonus fit for Croesus. You have to look sharp, play all the angles, keep talking to your brokers on the floor, your customers, your colleagues. You wear a headset and hang three telephones around your neck and keep your eyes glued to the screens stacked on your desk. You yell out a constant patter of prices, news, quotes, contracts, orders. You scarf your meals while reading the trade papers and never lose sight of the red lights marching along the wall that announce opening prices in Singapore, gold in Zurich, and anything else that might queer the pitch on a good day's speculation.

Speculate and related words *horoscope, telescope,* and *spy* share a common derivation with *specula,* the Latin word for *watchtower.* From their aeries over LaSalle Street, Chicago's speculators are watching the globe. They are wired among themselves into a chat group with computer links and speed-dialers to six continents. They talk to one another all day about prices, and these prices reflect people's opinions about the nature of the world. The prices are a poll, a vote up or down on the future of the Mexican economy, Japanese trade, or American debt. All day the speculators keep a lookout on the babel of numbers that tell them hours or days before the rest of us read about it in the newspaper that South African gold miners are going on strike or that Saddam Hussein's trigger finger is itching again.

The rules of the game are simple: buy low, sell high. How do you do this? Apply the Black-Scholes pricing model. Chart the "neckline" on a market graph. Second-guess the Federal Reserve Board. Follow George Soros. *Be* George Soros. Go for the demographic sweet spots. Invest as a contrarian. Read corporate reports. Ignore corporate reports.

A battle rages between those who say the financial markets are theoretically impossible to beat and those who say, "Hey, look at me, I'm a billionaire." On one side are the Nobel laureates, ensconced in the University of Chicago Business School, who are renowned for developing equations describing "efficient," that is, unbeatable, markets. On the other side are the speculators who beat them year in, year out with techniques "proven" not to work. But what if there were a system that *did* work, a method for finding pattern in chaos, for predicting trillion-dollar markets that most people assume are random? The predictors who discover this Holy Grail will be lords of the realm. They will possess a wonder-working money machine capable of producing a sweet fountain of cash. They will be rich, famous, and free.

Bustling with three hundred employees, the Swiss Bank trading floor is a big room lit by a wall of arched windows looking down LaSalle Street. The room is dotted with twenty-five octagonal turrets. In these computer-laden cubbies sit groups of traders specialized in various markets—foreign exchange, government debt, commercial loans. Other pods are working interest rate swaps and options with exotic payoff mechanisms. These derivatives, with names like *touch*, *up and in*, and *down and out*, sound like recent discoveries in particle physics. The room is wired into the markets by an electronic ticker tape of red lights dashing along the wall, hundreds of glowing computers and scrolling monitors, forty news channels, and scores of open phone lines. There is also the roar that comes up through the floor when the action gets *really* hot in the T-bond pit.

Except for several Swiss bankers who prefer pinstripes, the look at SBC-O'Connor is suburban casual. The men wear penny loafers, chinos, and sports shirts with polo ponies on their breasts. The women wear slacks and blouses open at the neck. The smell is Calvin Klein with a touch of Elizabeth Taylor. Each trading pod has its own character, depending on the markets they play. Equities traders are smooth talkers who know how to get a date on the phone. Forex is a zoo. The traders are hopping up and down on their chairs. They flash their fingers at one another and yell at phone clerks around the world to buy, sell, straddle, strangle! The turret reaches fever pitch before the book gets passed to Singapore at 3:30 P.M.

"In thirty seconds we're going to buy marks."

"At seventy-four oh-seven, I buy Swiss."

"We have a D-mark fill at fifty-eight ninety-five."

The top trader on the currency desk begins cavorting like a coach at the touchdown goal. "Go! Go! Go!" he yells, as the traders around him give one final push to lock in a price before the action moves to Asia.

Next to the foreign exchange turret is the equities desk, which is hustling through its own big day of trading several million stocks and options on the Chicago, New York, Philadelphia, and San Francisco exchanges. A carrot-haired man wearing blue jeans and a red work shirt sits at a desk with three computer keyboards connected to black, white, and green screens. Across the screens run ticker tapes and news flashes from Reuters, Bloomberg, and dozens of other financial services. The trader riffles through stacks of multicolored crib sheets covered with Greek symbols. These identify various kinds of risk. He watches the light board to make sure the bond market doesn't blow up in his face. He fields orders barked from his boss and telephone calls from downstairs, while all the time speaking softly, in a flat mid-western accent, into a telephone headset that is connected to a speed-dialer with one hundred and twenty dedicated lines open to the world's major exchanges.

The carrot-haired trader actually has *two* bosses sitting next to him. One is yelling to hedge the Swiss Bank portfolio short. The other boss is a computer, a snappy little PC lit up in fuchsia and blue. This fanciful, fun-looking device occasionally makes a noise: *Beep! Beep!*

At every *Beep!* the trader spins in his seat and starts moving to hit the speed-dialer. He can tell from where the numbers pop up on the screen which futures market he is supposed to call. New York for oil. The Merc for currencies. London for European bonds. And he can tell from the color—blue for buy, red for sell—what the machine wants him to do. In seconds, the trader is talking to a Swiss Bank clerk who is flashing the order into a pit. The order gets filled, marked up, time-stamped, and reported, usually within the minute, and never more than three minutes from the first *Beep!*

The Swiss Bank trader calls the computer at his elbow a black box. In the financial world, a box is a computer, and a black box is a computer whose program is a mystery to the uninitiated. The computer's signals are clear, but how it gets these signals—which sometimes contradict his other boss—is less clear.

"The magic gadget is a little threatening," confesses the trader. The box is emotionless, opaque, obscure. It gives no winks and nods. It has no voice, save for its unvarying *Beep!* But everyone on the floor is impressed by the box's one salient feature. It appears to have an uncanny knack for being on the right side of trades.

The black box is connected to a leased, high-bandwidth telephone line open twenty-four hours a day. The line runs from LaSalle Street west across the Great Plains and over the Rocky Mountains to Santa Fe, New Mexico. This old Spanish town is better known for being home to Kit Carson and Billy the Kid than for being a big player in the world financial markets. So who in the Rio Grande valley operates a computer flashing orders through a Swiss bank in Chicago to financial markets around the world? Has someone found the Holy Grail of market prediction?

Clairvoyant Trader

The reaction of one man can be forecast by no known mathematics; the reaction of a billion is something else again.

—ISAAC ASIMOV

Philosophers can easily be rich if they like, but their ambition is of another sort.

—ARISTOTLE

July and August are the rainy season in New Mexico, when the mountains green up and the sky fills with puffy white clouds. Doyne Farmer wakes to a red sun slanting over the Pojoaque River. Last week it was a dry-wash creek filled with sand. Today it is a broad sheet of water flowing through the canyonlands north of Santa Fe. He folds down the top on his Datsun 2000 roadster and sits next to Clara, the family dog. They roar south down the Rio Grande valley. Towering to the left are the cobalt peaks of the Sangre de Cristo mountains. Rising to the right is the yellow face of the Jemez Range. They pull into downtown Santa Fe and park behind an adobe bungalow near the plaza. It is August 1991.

Tony Begg greets them at the door. "Have you heard? The markets are crashing. The Dow is down seventy points."

"What's happening?" Doyne asks.

"Gorbachev is under house arrest, and George Bush, for the first time in his presidency, has canceled a golf game."

Doyne shrugs. "Unless you have a CIA mole on the ground, no one can predict a Russian coup." They enter the one-story house, which is furnished with plastic-webbed lawn chairs and folding tables holding five top-of-the-line Sun workstations.

"One night when I was monitoring currency flows, I noticed money flooding out of the Middle East," says Tony, in his clipped British accent. "The next day Iraq invaded Kuwait. The markets sometimes know the news before it becomes news."

They are joined in the living room by fellow researcher John Gibson, a tall young man with curly blond hair and a chin beard. Recently graduated from St. John's College in Santa Fe, which is famous for its Great Books program, John knows more about reading Euclid in the original Greek than he does about Wall Street. "So why would the Dow drop seventy points because Gorbachev is under house arrest?" he asks.

"Markets respond to news, and good traders make money because they supposedly know what the news means, but it's sometimes hard to believe," says Doyne.

John sips a cup of coffee. "It's amazing how fast the price moves when things heat up."

"That's why stop-loss orders don't always work," says Tony. "By the time your order gets executed, the price may be a lot lower than where you wanted to get out. When the markets are in free fall, you sometimes have no choice but to fall with them."

"That reminds me," says Doyne, "stop-loss orders are another thing we need to take a look at." He is keeping a mental list of topics they need to study. The list is already long enough to occupy them for months, but task number one is to develop models—quick and dirty, but effective models—for beating financial markets; demonstrate they work; and then shop for a partner with a spare hundred million dollars to invest. It is a tough problem, some would say an insoluble problem, but the eight people working in the little bungalow at 123 Griffin Street are confident they can crack it.

Their company has no name, no furniture, no money. But assembled on Griffin Street are some of the smartest people in the new science of chaos or complex systems or whatever one calls this branch of knowledge good at finding order within disorder. They are experts at predicting the trajectory of roulette balls, dripping faucets, and measles epidemics. Financial markets are next. Like any other dynamical system, these markets generate data points moving through time. Map these points in multidimensional universes and they reveal, if only for a fleeting instant, where the system will be next month, next week, or in the next five minutes. A clairvoyant trader betting on this knowledge could walk away from the table with a lot of money. How much money? Take the initial hundred-million-dollar investment, leverage it by a factor of five, secure a thirty-percent return, collect twenty-five percent of this as a fee for your services, and pocket a cool thirty-seven million dollars a year in profit. This is a nice balance sheet for Year One. After that, real money.

Doyne and his long-time friend Norman Packard have already written some of the key papers on how a system like this would work. They published their papers and went about their business, until the idea of beating the markets became irresistible. Doyne quit his job as head of the Complex Systems Group at Los Alamos National Laboratory. Norman quit *his* job as a tenured professor at the University of Illinois, where he was a member of the Center for Complex Systems Research. They hired two graduate students, a postdoc, and two research assistants and moved into an old adobe house between the Green Dragon Chinese herb store and the Santa Fe County court house.

On July 1, 1991, Doyne answered the doorbell at the Science Hut, as they call it, to accept express delivery of five Sun computers, costing about four thousand dollars apiece, which he gingerly placed on the floor. His next purchase was the folding tables and plastic lawn chairs that allowed this team of Ph.D. safecrackers to get to work.

With standard dress being rubber rafting sandals, Patagonia shorts, T-shirts, and long hair pulled into ponytails, 123 Griffin Street looks more like a graduate student lounge than a business seeking high-net-worth investors. The only real furniture is a big wooden table surrounded by mission-style chairs which fill the living room. These were purchased at the insistence of James McGill, the company's president. "You can't run a business off folding tables," he declared. "The whole operation looks like it could be gone in a minute."

After hearing Tony's news about the Russian coup, Doyne installs himself at the big table. His office, a small sun porch off the back of the house, still has no furniture, and the company's one telephone—a rotary model—is here in the living room.

"Tony, would you mind phoning Sun and reminding them to ship the rest of our computers?" he asks.

"I already have," Tony replies, clearing his throat. "They're a bit concerned about our credit rating, and the last time I looked, we had one thousand four hundred dollars in our account."

"That reminds me, I better call the bank. They turned down my credit application."

Doyne dials the number and begins explaining his situation. "I bought a house. That's really the only time I've ever borrowed any money."

Joe Breeden, one of Norman's Illinois graduate students, walks into the room and starts describing *his* financial problems. "You wouldn't believe the trouble I had last week when I first got here." Joe is wearing blue jeans with a hole in the knee and a long-sleeved shirt rolled up at the sleeves.

"I'm trying to rent an apartment. 'Where do you work?' they ask. I answer, 'The company doesn't have a name yet.' 'What's your salary?' 'I don't have one yet.' When I start telling them about my stock options and back-end payments they look at me like I'm a nut."

Joe reports that he was awake most of the night massaging eleven years' worth of data on the British pound. "I got the model up to a twenty-one percent gain. That's the real figure. I didn't leverage it," he says. Leveraged investments are made with borrowed money, which can significantly increase their returns. Speculators who can buy an options contract at five percent of its face value will multiply their potential winnings—or losses—twentyfold.

The next team member to walk through the door is Stephen Eubank. With a big red beard and bushy hair, a florid face, and blue work shirt rolled up at the sleeves, Stephen looks like a mountain man visiting town for provisions. He is actually a University of Texas particle physicist and Los Alamos expert on signal processing. "Anybody make any money yet?" he asks.

Behind Eubank is the barefoot Tom Meyer, who is wearing gym shorts and a fluorescent-yellow sports watch. An avid volleyball player and equally avid sports bettor, Meyer is grumpy from having stayed up too late at a bar on Cerrilos Road, which he says is the best place in town for watching games. Tom ducks into the little alcove in front of Doyne's sun porch and pulls a lawn chair up to his computer. It is not a good match: the chair is wobbly and Meyer is six feet seven. He starts punching in numbers and pushing the RUN button on programs meant to resolve millions of data points into hitherto unforeseen patterns.

Like an alchemist at his alembic, mixing eye of toad and tongue of newt, Tom is sifting through a mess of yen-dollar cross-rates, central bank statistics, and ticker-tape data. He is practicing the fine art of time series analysis. A time series is a set of numbers collected at various intervals. Some of the oldest time series, dating from the Renaissance, measure sunspots or flickering stars. But anything of duration, from a Mozart quartet to the stock market, can be reduced to numbers and analyzed as a time series. Into the computer go bit strings of 1s and 0s. Out come bar charts, moving averages, state space diagrams, and other draw-by-number pictures that allow even innumerate bystanders to say, "Here's a blip that keeps reappearing at regular intervals. Here it comes again, right . . . *now*."

A decade before they opened shop on Griffin Street, Norman Packard, Jim Crutchfield, Doyne Farmer, and Rob Shaw—the four members of the Dynamical Systems Collective, also known as the Chaos Cabal—published a key paper entitled "Geometry from a Time Series." This early cornerstone in the development of chaos theory showed that one could take a dynamical system, like water flowing through a pipe, insert a probe that measured what the fluid was doing at one point in time, and from this single probe reconstruct the behavior, or "geometry," of all the water in the pipe. This reconstructed behavior was displayed in something called a state space diagram, and no one was more agile at drawing these pictures than the Chaos Cabal.

"The movement of money in markets resembles turbulent fluid flow," Doyne explains. "There is a high degree of randomness, but there are also deterministic patterns that ride on top of this noise and give an overall shape to the market."

Jenny Cocq, a pretty young woman who wears her blond hair in plaits, walks through the front door, which is guarded by a Day-Glo picture of Albert Einstein smoking his pipe. Jenny is the office manager, who began her job by buying a chair to sit in. Accompanying her is Helen Lyons, their accountant. She has come to balance the books, which consist, at the moment, of Doyne's checkbook and a shoe box full of receipts. Helen, who also works as pastry chef and co-owner of Café Fritz, seldom arrives empty-handed on Griffin Street. Today she is carrying a chocolate cake with raspberry cream filling, which she places on the living room table. Long before lunch the cake will have disappeared.

Behind Jenny and Helen comes Jim McGill, who is wearing blue jeans and track shoes. He is the tough nut who is supposed to get this creative riffraff assembled into a bankable firm. His tonsured head, round face, and wire-rim spectacles give him the air of a monk. McGill is a physicist who jumped from academia into business. He was a guitar picker and laid-back surfer when he was in graduate school with Doyne and Norman at the University of California at Santa Cruz, but soon after graduating in 1974, he began retooling himself as an entrepreneur good at founding high-tech companies and taking them public.

As Tony wrote of McGill in an e-mail message describing their first meeting, "He's as laid back as the rest of us. Where's the sobering influence? But when we got down to business, his sharpness came to the fore, which is probably what we need."

McGill has no office, the house being too small to give everyone a desk. So he sets up his PowerBook in the living room and starts sending out a flurry of electronic messages and faxes. For confidential phone calls, he drags the telephone onto Doyne's sun porch and sits on the floor.

"What's this?" asks John, admiring some brightly colored graphs on McGill's computer screen.

"Portfolio management tools," he answers. "I'm trying to build a cash flow model for the company."

"Cash flow? That would be nice," says John. While finishing college, he used to work next door at the Green Dragon Chinese herb store, where he swept the floors and rang up the occasional sale. This is what he knows about cash flow, and it doesn't resemble what's happening here.

The next person to walk through the door is Norman Packard. Tall and amiable under a helmet of blond hair, he looks a bit rumpled after arriving on the night flight from Italy. Norman has been telecommuting by e-mail from Milan, developing models and chatting with his office mates, but this is his first visit to their corporate headquarters. He flashes his snaggle-toothed smile and says, "Cool, very cool." He catches up on everyone's news and then nips out for the first of the day's many espressos.

Back at Griffin Street, Norman sits in front of Seldon, his computer, which is named, like the other computers in the Science Hut, after a character in Isaac Asimov's *Foundation Trilogy*. The trilogy tells the story of "the last great scientist," Hari Seldon, who "could foretell the future mathematically." Seldon managed this feat through "psychohistory—the science of human behavior reduced to mathematical equations."

"Psychohistory is what we're up to," Norman jokes. "We agree with Asimov. 'Order must underlie everything, however disorderly it may appear.' "

Norman shares his office, which is really nothing more than an alcove located between the living room and Doyne's sun porch, with Tom Meyer. They sit in their lawn chairs with their knees up to their chins and their backs nearly touching. Norman begins loading a model for forecasting the British pound, so-named because it once equaled a pound of silver. "When do the London markets open?" he asks.

"I don't know," Tom replies curtly. "That's Tony's department."

Norman unwinds himself from his chair and crosses the living room into Tony's office. "When do the London markets open?" he asks.

"Six hours before Chicago," Tony answers.

"No wonder my model was so good. It was contaminated with future data. I was using opening prices in Chicago to predict the London opening."

"We've all made the same mistake," Tony confesses. "This will help you," he says, handing Norman a dog-eared copy of *The Wall Street Journal Guide to Understanding Money & Markets*. "Here's a nice diagram explaining when the markets open and close."

In his early forties, with a sprinkling of gray in his dark hair, Tony is the oldest researcher on Griffin Street. He is also the only one who has ever actually invested in the markets. After studying physics at Cambridge University, he was working as a computer programmer in London when he sold his house at the peak of the British real estate boom in the early 1980s. He invested his profit, sixty thousand pounds, in the stock market, where half the money disappeared in the crash of 1987. "I got depressed and broke up with my girlfriend," he remembers. "Then I decided, *don't get mad, get even.*"

Tony took his remaining money and started playing the futures markets, where he learned how to hedge portfolios by switching between Japanese yen and German marks. After reading their scientific papers, he struck up an e-mail correspondence with Doyne and Norman. He flew to meet them in Santa Fe, started playing his system on the Los Alamos computers, and then moved to Griffin Street, where he was supposed to produce an automated prediction engine.

Tom is staring intently at a screen full of digits as Norman walks back into their office. "We're not rich yet, but I see some nice numbers here." He stretches his arms over his head and emits a big yawn. Then his chair collapses under him. "Worthless piece of shit!" he yells from the floor.

Tom's irritability is due, in part, to a late night spent tweaking a computer program called Prophet. Prophet is Tom's name for an algorithm that was developed by Norman, his dissertation adviser. Prophet is a genetic algorithm, a program specially deigned for finding patterns in data. It mimics biological evolution by maintaining an evolving population of hypotheses, each good at locating structure in data and each encoded by a "genome" that mutates in various ways as the population changes.

Norman developed this program to analyze the "evolution" of snowflakes, but its first practical application was the study of Italian politics. Norman and his wife, Grazia Peduzzi, were living near her family in Milan in 1990, when Norman, in exchange for an office borrowed from his brother-in-law, agreed to analyze some data collected from seven hundred governmental offices in the Lombardy region of northern Italy. Norman's learning algorithm, which gets smarter at each pass through data, did an excellent job of figuring out why some Italian municipalities are more efficient than others.

The program was then borrowed by Tom. He was using it to study fluid flows—his dissertation topic—when one night he changed the inputs. He took out the wave data and replaced them with football statistics. Ten years' worth of data produced some surprising rules. When, for example, a visiting team is up against a divisional opponent it has lost to at home in the previous same-season matchup, and the Las Vegas line (the difference that the Las Vegas bookies expect between the points scored by the two teams) has it pegged as an underdog by at least three points, then bet on the visiting team. Meyer, his brothers, and some friends in Las Vegas began using Prophet to run a sports betting business. The operation cleared more than fifty thousand dollars in its first year. Tom used the money to buy a thirteen-bedroom former sorority near the University of Illinois, which he renamed Feynman House and rented out to his fellow graduate students. After moving to Santa Fe, Tom was still getting calls from tough-sounding characters in Las Vegas, until Jim McGill, fearing conflict of interest, made him promise to get out of the sports betting business. "Football or finance, all you do is plug in different data," says Tom.

Before leaving the office for his regular midday volleyball game, Tom hands Doyne a report on Prophet's latest run through the currency data. "Look at this," he says, shoving forward a graph of n-dimensional boxes floating in state space.

"What am I looking at?" Doyne asks.

"Futures data on the French franc, the British pound, and the German mark."

"What's this rising curve?"

"A fourteen-percent unleveraged advantage. I can do better, but it's a start. The best human traders are correct only forty percent of the time. It's just that they make more money when they're right than they lose when they're wrong."

"What years are you looking at?" Doyne is still trying to orient himself among the numbers. "We have to be careful. Some historical events, like the crash of '87 or German reunification, may have altered the nature of the currency markets."

Tom hands Doyne another graph with surprisingly good results. "My learning algorithm wasn't learning fast enough," he admits. "So I started teaching it things."

"I hope it didn't learn to overfit the data," Doyne says lightly. He is referring to a cardinal sin in the new science of nonlinear prediction. Linear equations, when graphically plotted, produce straight lines. Nonlinear equations produce curvy lines. "Life is nonlinear, and so is just about everything else of interest," remarked physicist Heinz Pagels. Doyne and his colleagues have spent years honing the mathematical skills required to analyze wiggly lines. They can derive equations for mapping moving points in two dimensions, or five or twenty. These equations can represent a vast array of data. The trouble is that some of the data are irrelevant to the patterns one is looking for— they are noise not signal. Graphing every available data point will tell you where you have been, but not necessarily where you are going.

On learning that Prophet is calibrated to make its predictions one day in advance, Doyne launches another round of questions. "We need to get transaction costs figured into our models," he advises. "Short-term trading strategies look good, but the cost of executing them may eat us alive."

"Tony, what's the slippage on the British pound?" he yells into the next room. *Slippage* refers to the cost of trading, primarily the bid-ask spread. This is the gap in price between what brokers charge when they are buying or selling contracts. The bigger the gap, the higher the cost of trading. "Is our slippage one tick? Is it five? Is it ten?"

"It seems to vary from market to market and what time of day it is and all sorts of other things over which we have no control," Tony replies.

"What's a tick?" asks the ever attentive John, as he works at his computer. He and Tony amble into the living room as Doyne begins explaining how the name comes from the old ticker tape. This was a one-inch-wide strip of paper that spewed steadily from telegraph printers that ticked as they ran. For over a century this tape, listing the price and size of every stock transaction, was the trader's bible. Computerized quote machines have replaced ticker tapes, but a *tick* still refers to the smallest unit by which exchange-traded contracts move up or down in price. *Tick data* is the register of every transaction made in a financial market. The new science of prediction is voracious in its appetite for tick data, because this is the raw material from which its learning algorithms learn.

Joe and Stephen wander into the living room, and Joe begins sketching his new model on the company whiteboard. It is based on time series analysis and other techniques in nonlinear forecasting. "It still looks as good today as it did last night," he reports.

Doyne seizes the ink pen and begins elaborating a model of his own. "I'm working on a theory of how our strategies can 'vote' together in an optimal way," he explains. "By combining Prophet with Joe's currency model and Tony's trading rules we should get a fourfold improvement in performance." When Doyne gets excited, ideas fly around him like startled pigeons. His voice rises. He pumps his arms up and down and loses himself in a speculative flurry.

Attracted by the sound of Doyne in full thought, Norman walks into the room and glances at Tom's graphs. "What's the Sharpe ratio on these predictions?" he asks. Sharpe ratios are the gods of modern investing; no one makes a move without consulting them. Named after Stanford economist William Sharpe, these ratios measure return-to-risk. A stock market system capable of making a big killing is useless if it leaves one vulnerable to going bankrupt. Everyone in the room knows they need a Sharpe ratio over 2.0. This characterizes investments with high returns and little risk of going belly up in what would otherwise have been a brilliant year.

Tony suggests they hedge their bets with options and other risk-limiting strategies good for pumping up Sharpe ratios.

"I'm still a little confused about how options work," John admits.

Now it is Tony's turn to step to the whiteboard for an impromptu Science Hut seminar. He launches into an explanation of how a call option gives you the right to buy a financial contract at some future price. A put option gives you the right to sell it. He takes over the ink pen and begins diagramming "the relationship between option strike prices and currencies as they drift in or out of the money."

"We can fine-tune this stuff later," interjects Jim McGill. "What we need right now are some numbers we can take out the door."

In *Gulliver's Travels*, Jonathan Swift describes the Grand Academy of Lagado, which is "filled with globes and spheres and mathematical instruments of all kinds." Here Lemuel Gulliver finds scientists whose "heads were all inclined either to the right or the left; one of their eyes turned inward, and the other directly up to the zenith." Intent on perfecting systems for extracting sunbeams from cucumbers or building houses from the roof down, the Academicians would have fallen over cliffs or done other bodily harm to themselves if they had not been protected by flappers. Flappers are servants who carry dried animal bladders filled with peas, which are rattled as an attention-getting device at particularly perilous moments. McGill is the Griffin Street flapper.

Sunlight from cucumbers. Money from the world financial markets. The schemes are fabulous. The practicalities sketchy. "The fact is, we don't know squat about what factors to use in predicting markets," Doyne confesses. "But there's a lot of voodoo in this business, and even the big shots have differing opinions about what's important."

How does it work? "The future, in some sense, will be like the past. Certain patterns have predictive value, because they replicate themselves through time. Our assignment is to find these patterns. This is science. The rest is voodoo."

McGill catches everyone's attention with another remark. "We can't start paying salaries until we incorporate ourselves as a legal entity, and we can't incorporate until we have a name. Maybe we should take a vote on what to call ourselves after tomorrow's board meeting."

Doyne hands out a piece of paper on which are written all the names that have been suggested for the company. The memo begins with a long list of synonyms for the word *foretell*. They include *predict, forecast, prophesy, divine, augur, portend, forebode, presage*. Other possible names include *oracle, prophet, nostradamus, Science Hut, Griffin Street forecasters, infinite regress, ETR* (which stands for "eat the rich" or "economic theoretical research"), *clairvoyant comrades, the vision thing*, and *dukes of destiny*.

As the meeting breaks up for lunch, Doyne reminds everyone that he and Norman will soon be going on the road. "We'll be doing a dog and pony act, looking for capital. So we have to push on getting a winning model, something with confidence levels we can sell. We need a lead pipe cinch."

"What's a lead pipe cinch?" asks Tony, who is unfamiliar with this cowboy method for saddling horses.

"It's equivalent to a royal flush in poker," Doyne explains. "No matter what happens, you win."

Equity

There's no doubt that money is to fore now. It is the romance, the poetry of our age. It's the thing that chiefly strikes the imagination.

—WILLIAM DEAN HOWELLS

Money talks, they say, but in what language?

—HENRY MILLER

Jim McGill is up early the following morning, sitting in the Science Hut, fiddling on his computer with cash-flow forecasts, while all around him, sitting at their computers, are researchers massaging market data into clairvoyant traders—computer programs capable of looking into the future and foretelling it. The office printer, balanced on top of a cardboard box, is whisking out a steady stream of graphs, histograms, research reports, and other evidence that the world financial markets are about to be cracked.

At nine o'clock sharp, a large black Cadillac pulls into the alley behind the house and parks next to the garbage cans. Out the driver's-side door comes a folding wheelchair with a portable telephone attached to the arm. Into the chair slides a brown-haired man with gold spectacles. James Pelkey, potential investor, has arrived, and the still-unnamed company is about to hold its first board meeting.

Pelkey rolls himself up the handicap ramp and slides from his wheelchair onto an old couch imported for the occasion. He slips off his shoes and uses his arms to lift his legs onto the couch.

"I guess we're all here," McGill announces, peering at the agenda on his computer screen. He is wearing his usual running shoes and jeans, but today his black T-shirt has two Mont Blanc pens stuck in the pocket. With his tonsured head and fringe of silver hair falling over his ears, he looks like someone who has taken vows in a financial order.

Seated around the living room in lawn chairs is a colorful assortment of gamblers and Ph.Ds. Doyne Farmer, his kinky brown curls beginning to gray at the temples, wears Birkenstocks, khaki shorts, and a banana slug T-shirt—the banana slug being the school mascot at UC Santa Cruz. Norman Packard, dressed in moccasins and blue jeans, has the kind of lanky grace suitable for starring in spaghetti westerns. Beside Norman is the sober, dark-haired Joe Breeden, who looks a bit wary, as if he is trying to avoid getting a fast one pulled on him.

Next to Breeden is the irrepressible Tony Begg, who is armed with a stack of diagrams showing "channel breakouts" and other prime moments in market timing. John Gibson, who bears a passing resemblance to the young John Wayne, sits at attention, his research notebook open on his lap. Off to the side of the room are the mild-mannered Stephen Eubank, with his flaming red beard, and Tom Meyer; wearing batik shorts and a sleeveless shirt, Tom looks stripped for action.

"As soon as we find a name, we'll get the lawyers involved in drawing up a partnership agreement," McGill begins, in a monochromatic voice, bleached of emotion, flat as a prairie. "Here's what I've drawn up as a list of Q3 objectives."

"Excuse me . . ." John seems to be resisting the urge to raise his hand. "What does 'Q3' mean?"

"Third quarter," says McGill, not missing a beat. "It ends the last day of September." He returns to his list.

"We have to incorporate, which means we need a name for this beast. I've hired Gordie Davidson at Fenwick & West in Palo Alto to draw up the papers. He's the guy who incorporated Apple. He's the king of venture capital lawyers.

"We need a company account, so we can stop running the business from Doyne's checkbook. We need a Q4 operating plan. We'll have rolling objectives every quarter. You rank your objectives and grade yourself on your performance," he explains.

"This sounds funny to me," Joe says. "I mean, we're doing research here. We're not building TV sets."

McGill is unmoved. "Like everyone else, we're trying to get a product out the door. As far as I'm concerned, *plan* is a verb."

McGill has borrowed this idea of corporate objectives from David Packard, Norman's cousin, who teamed up with his college friend Bill Hewlett to found the electronics firm Hewlett-Packard. The "H-P Way," as Packard called this idea of corporate objectives, is now standard practice throughout Silicon Valley and even in the federal government, where Packard served briefly as deputy secretary of defense.

"We want to move as fast as we can toward an experimental release and field trial. We might want to set up a ten-thousand-dollar fund inside the company and do some trading to shake out the bugs. Then we can write up the code and get a 1.0 release ready for production."

"We need a real-time data feed," Norman says. "We can't play the markets with day-old numbers printed in the newspaper." Everybody agrees. Getting market data piped into Griffin Street is a high priority.

"I've drawn up a three-phase business model," McGill continues. "It's a map for getting ourselves on the road. We begin by defining the business. Do we sell software? There's a company in California selling arbitrage software for a million dollars a copy. But neither Pelkey nor I think we should be selling software. We don't want the market to discount this technology.

"Do we get into money management? In this case, we have to raise between two hundred million and four hundred million dollars. Pelkey thinks money management is the way to go." McGill nods toward his longtime friend, who is sitting with his arms crossed over his chest.

"I favor partnering," McGill goes on. "We become an extension of someone else's research department, but keep a cut of the profits for ourselves."

"What comes next?" Tony asks.

"Once the company figures out what it's selling, we grow in three phases. In the start-up phase, the senior founders, that is, Farmer, Packard, and McGill, will receive no salaries, while everyone else gets by with graduate student wages. Starting in September, we'll burn twenty-five thousand dollars a month. We need one hundred thousand dollars through the end of the year and two hundred fifty thousand dollars in the first six months of 1992. If we have to go out and raise this money, it will be expensive. We'll have to give away equity to get it.

"Here's what I expect will happen as we ramp up to Phase II, the partnership phase. We stop selling stock in the company and start receiving payment and royalties. Phase II is about getting a positive cash flow. When everyone is receiving full salaries, I estimate we'll need a million dollars a year in revenue for the company to break even, and I hope this money will come from contracts with trading partners.

"In Phase III, when our forecasting technology is fully online, we move to raising our own funds and managing them," he concludes. "The company is so rich that we no longer need to work for anyone but ourselves."

"I'm for moving straight to Phase III," says Norman. There is laughter around the table as he hastens to explain. "We shouldn't get locked into supporting clients, if we can launch ourselves quickly into managing our own funds."

Everyone in the room agrees. As soon as possible, they should press on to the El Dorado of Phase III.

"Here's a diagram of our corporate structure," McGill begins, handing out copies of the document drawn up by Fenwick & West. The diagram shows interlocking circles of shareholders, senior founders, founders, board members, corporate officers, and employees assembled into something called an S corporation. "An S corporation provides an insulating shell around us in case we get sued," he explains.

Norman and Doyne, who are founders and part owners of another company called Eudaemonic Enterprises, ask McGill a lot of questions about how these interlocking circles relate to one another.

Eudaemonic Enterprises, whose sole product was a couple of toe-operated computers that were built into shoes and used to beat the game of roulette, was organized—or disorganized—in a very different fashion. Ownership in the company was strictly egalitarian. Anyone who donated labor, money, or ideas to the project received a slice of something called the Eudaemonic Pie. The pie, consisting of winnings plucked off the roulette tables in Las Vegas, was occasionally sliced up and served to pie holders in proportionate pieces.

"This looks like a Rube Goldberg machine," Doyne complains. "I'm nervous about the hierarchical implications in this diagram. I want to beat the system, not join it."

McGill presses on with the next item on his agenda—the budget. "It's like a flight plan," he explains. "It doesn't guarantee you'll land in Philadelphia, but it shows you how to get there."

"This is particularly interesting to me," says Doyne, "since the money is coming out of my bank account." So far he has spent about twenty-five thousand dollars on computers and other company expenses. Pelkey has chipped in fifteen thousand dollars, and Tony has put up another ten thousand dollars.

McGill starts going over the numbers. "Looming expenses include ten thousand dollars to set up an S corporation in Delaware and a burn rate at full salaries of ninety thousand dollars a month. Even a bare-bones budget, with the partners subsisting on twenty-five hundred dollars a month, will consume one hundred thousand dollars by the end of the year."

"Why don't we suspend salaries for another month," Joe suggests. "This will keep expenditures down."

"I could do without a salary," John agrees. "I understand money is very expensive right now."

"We're not counting on hiring any new people this year, are we?" Tom asks.

"No, we're going to be firing them," says Norman. Everyone laughs.

"I'm having a problem projecting revenues," McGill confesses. "It would help if we knew what we were selling. That's where you guys come into the picture. Forecasting, performance, risk, volatility—I need numbers to plug into the model. Then the numbers go into a sandwich. This way we can see what we're going to be eating for lunch."

"Lunch!" Tom exclaims. "That sounds like a great idea."

Tony produces the menu from a local restaurant and phones in the order.

McGill returns to his agenda. "If we're really lucky, we'll get the company bootstrapped for one hundred fifty thousand dollars. But it might take up to five hundred thousand dollars to get it launched. September was the date we promised to start paying salaries. I have to give Pelkey a financing document next week. This is a high priority, or there aren't going to be any salaries."

After emerging from his Cadillac, Pelkey has spent the morning stretched out on the couch, with his eyes closed, lying so still that one might have thought he was asleep. He wears a short-sleeved shirt, black trousers, and socks. A coffee cup is nestled between his legs. Next to him is the wheelchair and cellular phone that he picks up occasionally for incoming calls. The rest of the time he sits with his head down, motionless, except for the rare comment he shoots into the conversation.

Trained as an engineer, Pelkey is a Harvard MBA, Silicon Valley entrepreneur, and financier whose fast-track career skidded off track when his former wife put a bullet in his spine, which left him paralyzed from the waist down. During the first phase of his business life, Pelkey displayed a knack for attaching himself to start-up companies that later got sold for handsome prices. The most successful of these was Pacific Data Products, which made memory boards for laser printers. In 1989, three years after investing in the company, Pelkey and his partners cashed out for seventy-eight million dollars.

Pelkey was so good at sniffing out talent that Montgomery Securities, banker to Silicon Valley, made him a general partner in 1984 and put him in charge of a hundred-million-dollar investment fund. While Pelkey was handling their venture capital, Montgomery became one of the largest underwriters of small-company initial public offerings in the United States. The two Jims—Pelkey and McGill—have known each other since 1980, when Pelkey bought controlling interest in

Digital Sound, a company founded by McGill, and became its president. Ever since then their business lives have been intertwined, with Pelkey either investing in McGill's companies or serving on their boards of directors.

When Pelkey and his new wife, Dorothea, were thinking of moving to Santa Fe, Jim McGill introduced them to Doyne, who must have done a good job of extolling its virtues. Pelkey moved to Santa Fe in the spring of 1990. He is in constant physical pain, sometimes irascible, but always fiercely independent as he wheels himself around town.

McGill continues laying out the budget. "I want to see us making Wall Street money with a Silicon Valley model for starting up the company. What we'll be doing differently from Silicon Valley is blowing money out the other end to the partners."

In new technology companies, explains McGill, the venture capitalists who put up the money generally make the money. Company founders labor away at moderate wages, building up the business, until it gets sold by the investors, who walk away with most of the goods. "There are two ways to make money: cash flow and sellout," he explains. "In Silicon Valley you make money only on sellout, not on cash flow."

"Blowing money out the other end to the partners" is McGill's way of saying that he intends to give everyone in the room a portion of the company's profits, in the form of dividends. "You'll also own enough of the company to make it interesting, in case of a sell-out."

"How much of the company *will* we own?" asks Joe. This is the question that has kept him on the edge of his lawn chair for the past hour. The younger members of the team are willing to work long days at low pay, but they want equity. They don't just want to work for the company; they want to own it.

"Along with your equity stake in the company and its cash flow, you have to factor in salaries and bonuses. It's a whole package," explains McGill, who is trying to defuse what he suspects will be a volatile discussion.

"Talking about cash flow," Norman interrupts, "how are we going to pay for our sandwiches?" Doyne hunts around for his checkbook, pays the deliveryman, and starts handing out the lunch order.

"Bonuses are established from two criteria," McGill continues. "How well you meet your objectives and how well the company meets *its* objectives. It's a two-way street."

"Egos are already going to cause us big problems," Doyne interjects. "I think bonuses should be based on all of us pulling together."

"I'm just telling you how it's done in a normal company," says McGill. Then he turns to Pelkey and asks him to describe how bonuses were computed at Montgomery Securities.

"Our annual salaries were thirty-six thousand dollars, but the general partners took home between one and two million dollars a year," Pelkey says. After this cheery thought, his next comment is less reassuring. "Bonuses clip off the investors' share of the revenue flow, and the investors may want to see their money before the employees see their bonuses."

"So how do we tweak the bonus knob?" asks Norman.

"Usually ten percent of salaries is distributed as bonuses," says Pelkey.

"Tech Partners told me they were going to distribute twenty-five percent," Norman counters. Tech Partners is an experimental fund started by the big investment bank Crédit Suisse–First Boston. The firm had spent months courting Norman and his graduate students, trying to lure them to Wall Street.

"It's not impossible to get a bonus this big," says Pelkey. "But you have to remember that venture capitalists like to get a tenfold return on their investment. This compensates them for all the start-ups that fail."

"Will bonuses be distributed equally, or will they be proportional to our salary?" Tony asks.

"It's not standard practice to make them equal," says McGill.

"But I think they should be," counters Tony.

"How about Jenny?" asks McGill, talking about the company secretary, who is out buying file folders. "Are you going to give her the same twenty-thousand-dollar bonus you get?"

"Yes," Tony answers.

"We are arguing that virtue exists," says John. Apparently he agrees with Tony.

"I'm more interested in equity than bonuses," Stephen interjects. He is referring to equity in the financial, rather than the moral, sense.

While everyone around him is wolfing down avocado and cheese on whole wheat, McGill uses an overhead projector to beam onto the wall a picture of the company's "equity matrix."

"Jim Pelkey and I used our gut feelings to put the numbers in here. This assumes we need one hundred fifty thousand dollars in seed money through the end of the year and that we can raise the money at certain prices. This is the most expensive money."

"How expensive?" asks Norman.

"If the company is valued at two million dollars, this much money borrowed at this stage of the game gives ownership of seven point five percent." For the first time everyone in the room is alerted to the fact that Pelkey will own a good chunk of the company if he invests the initial capital.

McGill begins decoding the numbers in his equity matrix. "There are three senior founders and five founders. At an equity ratio of four-to-one, the senior founders, in the best-case scenario, will each own twenty-one point eighteen percent of the company. . . . "

"Wait a minute!" Tony interrupts. "By an equity ratio of four-to-one, does that mean you, Doyne, and Norman get four shares in the company for every one share that the rest of us get?"

McGill is staring at the numbers. "These are the standard ratios in Silicon Valley start-ups," he says.

Tom is blinking with disbelief. "So you guys are worth four times more than I am?"

Joe mutters, "When we go out to lunch, will we have a junior table and a senior table?" Even the mild-mannered Stephen looks visibly distressed.

The junior partners have launched what comes to be known as "The Revolt of the Masses." This slogan is scrawled in red ink on the company whiteboard the following morning, along with the maxim: "In the land of the blind, the one-eyed is king."

"What are the arguments for senior founders being worth four times more than junior founders?" Tony asks. The word "junior" appears nowhere in McGill's presentation. He distinguishes between "senior founders" and "founders," but now that Tony has attached it, the diminutive sticks.

"There are two bankable players on this team," McGill says flatly. He is referring to Doyne and Norman. "None of the rest of you could go out and raise money. Their bankability has to be rewarded. Otherwise, they might walk. The bad news, Tony, is that senior guys make more money than junior guys, and they get more stock."

Tony stands his ground. "This is our company, and it doesn't have to be this way."

"If you don't need money, you can do whatever you want," McGill comes back. "But since we need money, we have to look rational to the outside world. They are not going to finance a communist conspiracy. They don't want everyone to own the same amount of stock. They want to be sure the senior partners are locked in by their compensation plan."

McGill, looking for someone to back him up, turns to Pelkey. "What are the ownership ratios that ring warning bells?"

"Norman and Doyne are the big draws for the company," Pelkey confirms. "But you also have to have a working team. Everyone has to feel satisfied and good about it. That's part of the philosophy of this company. You don't want to come back in six months or a year and be unhappy about what you've agreed to."

Doyne, who has been fidgeting in his chair, finally speaks up. "I think we should go around the room and get a reading on how far apart we are. Maybe we should take everyone's answers and settle on the mean."

Tony fires the opening salvo. "I want the ratio near two-to-one. I can always pack up and leave for England if the deal doesn't seem fair. All I have is a Jeep lease that I can cancel on my way out of town."

Joe votes with Tony. "In exchange for owning more of the company, I'll settle for a subsistence salary and smaller bonus."

"I'm happy with a minimal salary, too," says John. "I'd rather have shares in the company. I have faith in it."

Stephen acknowledges that Doyne and Norman have a lot of work ahead of them attracting investors. "I vote for three-to-one."

"I was lured out here on the promise of earning eighty thousand dollars a year," says Tom. "I opt for a higher salary, even if it means fewer shares in the company. I'd like to buy a house and a car that starts. I agree the senior partners warrant more than the junior partners. I'm not excited about it, but I'll settle for four-to-one. I don't want to see the value of my shares watered down, though, as more of them are handed out over time."

McGill shakes his head. "This is called antidilution rights, and not even the venture capitalists get it."

"Yea, that's what I want," says Tom, "some antidilution rights."

"I think four-to-one is the right number," McGill declares. Then, his patience wearing thin, he says, "I'm walking a fine line between being CEO of the company, and being someone who merely helps to get it started."

Norman weighs in with an impassioned speech. "Doyne's and my fifteen years' experience in the prediction business should count for something. We have intrinsic value. People believe our results are valid because *we're* sitting here saying so. I've outgrown my idealistic phase. Communism is not the right paradigm on which to construct a company."

"Especially if you want capitalists as investors," McGill adds.

"At two-to-one I'd be inclined to run," Norman concludes. "The main argument against my running is that this is a group of people I care for."

"This is true for all of us," says Joe.

"I feel pretty schizophrenic about all this," Doyne confesses. "I'm hearing two voices in my head. One says I should be a good guy. The other says I want a fair share in the company. I'm doing this so that ten years from now I'll have enough money to found my own research institute. But what if it goes in the other direction? Norman and I will take the rap if this thing goes down. My career is a strange one, and our track record here means a great deal to me."

Then he looks around the room. "We need a compromise between two cultures," he says. "Jim is an animal coming in from a different world; so we have to treat him like he's used to being treated in that world. But at the same time we have to create an all-for-one atmosphere. The hierarchy sitting on top of us can't be perceived as oppressive."

"How important is equity in the first place?" Norman asks. "What if the value of this technology goes to zero?"

"I don't see it going to zero," Doyne replies. "I see the value spiraling upward at an exponential rate. Soon we'll be light-years ahead of what we're doing now, and everyone using this technology will be involved in an escalating battle."

Equity comes from *aequitas*, the Latin word for fairness. It is a legal term. It is also a business term, where equity refers to the market value of a company minus its debt. It is no accident that the word has both moral and financial implications. Dividing up ownership in a company involves dividing up responsibility for its success or failure. Who will control the company? The investors who lend their money? The seasoned founders on whose ideas and reputations the company is built? Or the junior partners who labor to implement these ideas?

What is the relative worth of the people sitting in this room? Without Doyne or Norman, no one is going to give these green physicists two hundred million dollars to manage. But without the five young turks, Doyne and Norman may not get the project launched. Who is going to design their models? Who is going to write the computer code to get them online? Who is going to wash the coffee cups? At what price does one value this labor? These questions about equity, be it fairness or stock options, are integral to the process of starting a company. Embedded in the business plan is morality. Entailed in the cash flow is justice.

Pelkey's authoritative voice rises from the couch. "I think four-to-one is the right ratio," he declares. "Whatever you do, don't go below three-to-one. The founders have worked on this for many years, and Jim's role will become more obvious to you in the weeks to come. You're driven to have a democratic, equitable, 'communist' approach to the company, but if you spend all your time in meetings like this you'll never get anything done."

"I'll defer to the senior partners," Tony offers. "We have to have faith in letting them make the decision, and it will be a good test of what we're doing here."

It is late in the afternoon when Pelkey slides into his wheelchair. "We want to raise as little money as possible," he says. "If we in this room can put up twenty-five thousand dollars in September, without going outside, then we'll be in good shape. I'm impressed by what I heard today," he declares, wheeling himself out the door. "So I think we ought to hold off on raising capital and keep the equity in this room."

Pelkey is in! He is good for the September salaries. They're launched. As his Cadillac rolls down the back alley, the room fills with cheers and whoops of delight.

Suddenly there is a knock on the front door. Tony opens it to find two men standing in front of a yellow truck. They look perplexed. "We're searching for an unnamed company on Griffin Street."

Doyne steps to the door. "You must be delivering my desk and chair." He leads them back to his sun porch and starts writing a check.

The interruption reminds them that they have one unfinished item on the agenda. "We need to find a name," says Tony.

"I move that we walk down the street and reconvene over a drink," Norman proposes. The motion is seconded and approved.

"We have an action plan," Norman tells Doyne, when he comes back in the room. "We're going to get a drink."

Zozobra

Gold, gold, what an excellent product. All wealth derives
from gold, and it's gold that mobilizes all human action.
— CHRISTOPHER COLUMBUS

Put all your eggs in one basket, and watch that basket.
— ANDREW CARNEGIE

After Pelkey's Cadillac has nudged past the garbage cans in the back
alley, everyone else files out the front door and strolls two blocks east
to the Santa Fe Plaza. The big public square, lined with false-fronted
buildings and wood-timbered porticos, is filled with its usual collection of stray dogs, travelers, white-turbaned Sikhs, unhorsed cowboys,
Japanese tourists, low riders, and Navajo silversmiths, who squat in
front of the Palace of the Governors with their wares spread in front
of them on blankets.

The researchers do a quarter turn around the plaza and walk up a
flight of stairs to the Ore House bar, which serves eighty different
kinds of margaritas. Norman orders a pitcher and joins everyone on
the balcony overlooking the village green. From down below come
the murmur of voices and the sound of pedestrians strolling the art
galleries, which have replaced the furriers and iron mongers that once
did good business here. They lean their chairs against the wall and
munch on tortilla chips and salsa as the alcohol delivers its first pleasant punch to the head.

They stare out over the adobe buildings of Santa Fe, which are plunked higgledy-piggledy on winding streets shaded by cottonwoods. Beyond the tile-roofed houses rise the ruby peaks of the Sangre de Cristo mountains, which give way below Santa Fe to the mesas and dry arroyos that stretch from here to the Mexican border. In the warm red glow of the afternoon, the city, tucked under its apron of trees, looks half asleep. It will not wake from its siesta until the hard yellow glare of the desert sun has mellowed into the blue tones of evening.

Except for the occasional car poking through town, the scene is little changed from the days when Indians, conquistadors, trappers, and traders raced their horses around the plaza to mark the end of a long ride up from Mexico City or across the plains from Saint Louis. Santa Fe was always the end of the trail, the Timbuktu of America, a fabled city of riches sequestered on the far side of a nearly impenetrable wilderness. The city's origins are linked to gold, romance, commerce, and the big payoff that always somehow managed to recede over the infinite horizon.

The first gold-crazed explorer to pass through town was Francisco Vásquez de Coronado. In 1540, eighty years before the Pilgrims set foot on Plymouth Rock, Coronado and his fellow conquistadors were searching New Mexico for the seven cities of Cíbola. The Cities of Gold turned out to be Zuni pueblos made of mud. Coronado retraced his steps down the Rio Grande and died a broken man, but his successors all had their own hopes of finding in Santa Fe a golden horde.

At seven thousand feet, Santa Fe lies alongside a stream that flows from the Sangre de Cristos southwest to the Rio Grande. This was a favorite hunting ground for the Pueblo Indians, but it was not permanently settled until 1609, when fifty Spanish soldiers built some mud and timber houses around the Plaza that for the next four centuries would be the center of Santa Fe life. In Coronado's wake came the priests and settlers who did the actual work of colonialism— converting the Indians, enslaving them, and murdering the less tractable among them. This civilizing mission was briefly interrupted in 1680, when twenty-five hundred Pueblo Indians, chanting songs of vengeance, barricaded a thousand hungry Spaniards, along with their equally hungry sheep, mules, horses, goats, and cattle, inside the Palace of the Governors, the block-long mud building that dominates the north side of the plaza.

The Indians had already killed several hundred people in the villages north of Santa Fe. They burned the church on the east side of the plaza and eventually expelled the Spaniards from New Mexico. They blocked up the windows to create traditional rooms, entered from the top by ladders, and converted the palace into a pueblo village. This revolutionary war of independence marks the sole occasion in the history of North America when indigenous people succeeded in removing their colonial overlords.

The Spanish licked their wounds for a dozen years, before Capt. Gen. Diego de Vargas Zapata Lujan Ponce de León y Contreras, armed with a willow-wood statue of the Virgin Mary, rode back into Santa Fe and reclaimed it for Spain. Every September the city reenacts this event in the great colonial bash known as Fiesta. Ladies dressed in mantillas and combs parade their horses through town. Beside them ride their caballero boyfriends dressed as conquistadors. Leading the procession, seated in a canopied palanquin, is de Vargas's madonna, Nuestra Señora de la Conquistadora—Our Lady of the Conquest—who wears a richly jeweled crown and silk mantilla sewn specially for the occasion. Everyone takes the parade very seriously, especially the actors chosen to reincarnate de Vargas and his queen, who also preside the following evening over the *gran baile de pasatiempo*, the fanciest of Santa Fe's fancy-dress balls.

Once it had resumed its colonial mission, the palace was occupied by a long line of Spanish governors, who were followed by territorial governors, after New Mexico was ceded to the United States in 1846. The most famous of these was Lew Wallace, who lived in Santa Fe in the late 1870s, while he was writing *Ben Hur* and fighting desperados like Billy the Kid. "I mean to ride into the Plaza at Santa Fe, hitch my horse in front of the Palace, and put a bullet through Lew Wallace," said the Kid one morning.

Before murdering twenty-one men—which averages one per year during his short life of crime—William H. Bonney, a.k.a. Billy the

Kid, washed dishes in the kitchen at the La Fonda Hotel. This Santa Fe landmark lies directly across the plaza from the Ore House. Sheriff Pat Garrett, after dispatching the pestiferous Billy, checked into the La Fonda to spend the night. The hotel is also famous for marking the end of the Santa Fe Trail, which began in Independence, Missouri, and terminated eight hundred miles later at the La Fonda bar.

The La Fonda began serving them in 1821, but before then, white people of Northern European descent were barred from Santa Fe. When he reached the city in the spring of 1807, after a miserable winter in the Colorado Rockies, during which he and his two dozen men nearly starved to death, Lt. Zebulon Montgomery Pike was arrested by a hundred mounted Spaniards. Pike, who expected to find a city full of palaces built from the spoils of the Chihuahua silver mines, was surprised to discover a "miserable" town of mud houses, where agriculture and commerce were "a century behind" the Yankee settlements back east.

Pike was interrogated by the Spanish governor, sent to Chihuahua for further investigation, and then expelled from the territory via Louisiana. The Spanish confiscated his maps and notes, but Pike managed to pen from memory an account of his travels, which gave Americans their first view of Hispanic life in New Mexico and their first hint that good money could be made from breaking the Spanish monopoly. Soon it was time for the Spanish to join the Indians in aristocratic disdain, as they watched invading peddlers do a victory lap around the plaza. The Anglos invaded not only the La Fonda bar, but also the side streets off the plaza, which were soon filled with gambling halls and what one Yankee trader described as the "relentlessly coquettish *mujeres* of Santa Fe." Mingling with the teamsters and the *señoritas* dancing fandangos were a motley crew of buffalo hunters, *vaqueros*, prospectors, and soldiers quaffing the local raw whiskey known as Taos Lightning.

Two hundred years after Pike opened it, the Santa Fe Trail is about to get linked into another commercial network. Thanks to the anonymity of the global economy, where money is turned into bits and bytes and flows around the world through satellite transponders and fiber-optic cables, no one in the financial markets will know that a new player has entered the game and started placing bets from an adobe bungalow off the Santa Fe Plaza. This remote outpost is about to get plugged into the light boards and speed-dialers linking markets in Tokyo, Singapore, Zurich, London, New York, and Chicago. Once their financial signals start flowing east to west and back again, the scientists, who, on this mellow night, are leaned back in their chairs on the Ore House porch, expect quite soon to be doing their own victory lap around the plaza.

They order another round of margaritas and move to discussing unfinished business. "I think we should consider Mayan names," says Doyne. "The Mayans were obsessed with time and cycles and predicting the future."

"The Greeks were also big on forecasting," John adds. "Predicting solar eclipses was one of the earliest and most dramatic advances in science. It would be great if we could sign our papers of incorporation during an eclipse, right at the moment of maximum coverage."

"Reading entrails, studying global financial flows—it's all the same thing," quips Tony.

"I'm for keeping it simple," says Norman. "The two names I like best are Griffin Street Forecasting and Metis."

"What's 'Metis'?" asks Tom.

"One of the Greek words for 'wisdom.'"

"That's nice," he says.

With the votes equally divided between Norman's two suggestions, they decide to break for the night and name themselves in the morning. Also left unsettled is the equity debate. Who is going to own this company?

Over six feet tall, Doyne Farmer is rangy and slightly akimbo, with various parts of him, like his nose and fingers, having been broken several times and not always put together straight. His angularity is accentuated when Doyne starts thinking. He gesticulates and paces, and his voice rises with the sheer enthusiasm of tracking an idea to ground.

But tonight it is a subdued Doyne who walks into the house he shares with his wife, Letty, and their two sons and daughter on the banks of the Pojoaque River, fifteen miles north of Santa Fe. He is depressed by the day's events. The research results are encouraging, and it is good news that Pelkey is onboard, but the equity fight depresses him. A new company should begin with shared ideals, with everyone pulling together to make sacrifices for the common good. This isn't a job. It isn't a postdoc or a fellowship. It is people pooling their ideas to make something new. It should be a charged, almost mystical experience, but already people are squabbling. The common vision is being eclipsed by self-interest.

Doyne drives across the river to pick up the children from the woman who cares for them after school. He lets Clara the dog run loose on the dirt road. He organizes a game of kick the can and lights the outdoor grill. Norman, his wife, Grazia, and their son, Daniele, born earlier that summer, are coming for dinner. Clara gives Letty a big *woof* when she drives up in her dusty little station wagon.

Blue-eyed and quick, with her own deep reserves of energy and intellect, Letty is returning from a day of doing the public good. She runs the Santa Fe office of a law firm that specializes in environmental cases, mining rights, and other hot-button issues. Today she was advising members of the Navajo Nation to settle a fight about logging the reservation, not by suing one another, but by sitting down for a traditional meeting of the elders. She wears a floral dress with a tight-fitting blue jacket. Her only jewelry is a turquoise wedding band. Her blond hair is cropped at uneven lengths around her face, shoulder-length in the back, shorter in the front. She is tall, fined-boned, angular, with a golden aura around her like a Socratic dialogue in motion.

She and Doyne have made a striking couple from the first day they met in college. The star athlete and one of the brightest students in her class, from St. Timothy's through college and law school at Stanford, Alletta d'Andelot Belin comes from a long line of public servants. The most illustrious, if misguided, of them was her uncle McGeorge Bundy, who advised President Kennedy to get into the Vietnam War. But Letty long ago eschewed the trappings of her Boston Brahman caste to adopt an unadorned, unpretentious, and radically democratic way of being in the world. She is radiant and calm, even-tempered and judicious—traits that Doyne, when agitated, sometimes lacks.

She listens to him recount the day's events. "When it comes to dividing up ownership in the company, you shouldn't sell yourself short," she advises. "This is your and Norman's idea, and it's only going to work if the two of you manage to get it off the ground."

"We're the bankable partners," Norman reiterates, as the conversation continues over dinner. Everyone is sitting outside at a picnic table overlooking the river. Above them, the black umbrella of the night sky is pierced with a million starlit holes, and a crescent moon is rising over the mountains. "When it comes to forging corporate alliances or putting together an investment fund, you and I will be chasing the money. We need McGill for the business end. We need the junior partners for coding the programs. But it's us they want to buy.

"I hate to say it, but you, McGill, and I are worth a lot more than the junior people," Norman continues. "McGill is a seasoned pro. You and I have been in the prediction business a long time. I know you want to be democratic and make everyone feel good," he says to his old friend, "but McGill is going to walk if we don't give him a big enough share in the company."

The silent partner at the table, but whose opinion they already know, is Norman's wife, Grazia Peduzzi. While Norman is tall and laconic, with easy-going Western ways, Grazia is diminutive and intense. She is dark-haired, with an aquiline nose and piercing eyes. She dresses with the flair of someone who grew up in a family of Italian couturiers, which she did. She is a philosophy teacher from Milan, impassioned about politics, art, and the dangers of Norman selling himself short.

Grazia was present during the roulette project in the late 1970s, when a dedicated team of physicists and friends, who were living together in a big house in Santa Cruz, spent years building computers into their shoes and commuting to Las Vegas to break the bank. It was a communal effort, mounted by the company called Eudaemonic Enterprises. Labor, ideas, time, money—everything invested in the project went into a central accounting system, called the Eudaemonic Pie. When plump with roulette winnings, the pie was to be sliced up and served in equitable portions.

While Grazia taught Italian at the university, she watched Norman build toe-operated computers, instead of doing more practical things, like finishing his Ph.D. At the time, Doyne, Norman, Grazia, and numerous other people who came and went over the years were living together in the big house that Letty had bought on Riverside Street, near the Santa Cruz boardwalk. Back when they were friends and not lovers, Norman tutored Grazia in English and shared her passion for music. Norman is an amateur pianist and singer of Renaissance motets. Grazia is also a singer, with a rich contralto trained by the Church. Later she switched to singing folk songs at mass rallies devoted to organizing factory workers. Today she prefers opera.

Everyone at the table knows that Grazia had been plugging for Norman to take the Wall Street job at Tech Partners. This boutique investment firm, founded by Bill Cook and other runaway bankers from Morgan Stanley, is being financed from the deep pockets of Crédit Suisse–First Boston. Tech Partners had promised Norman eight percent of the company and a nice six-figure salary. While Grazia supports his leaving the university for Wall Street, she is dubious about Norman's going to work for an unnamed company in New Mexico with negative assets. She also has a low opinion of Santa Fe, which reminds her of a provincial Italian hill town, a nice place to visit, but no place to live. As Grazia sits at the table, tight-lipped and sober, her emotional radar is warning her that Norman and Doyne are about to get in trouble again.

They have been getting in trouble together for a long time. They grew up in Silver City, an old mining town in the red desert country of southwestern New Mexico. If one follows the Rio Grande south from Santa Fe to the last spur of the Rocky Mountains, there, on the southern edge of the Gila wilderness, at six thousand feet, sits this little town of twelve thousand souls. Towering above Silver City are the Ponderosa pine and Douglas fir forests of the Mogollon Mountains. Falling away below it are the lava beds and salt flats of the Sonoran Desert. At its historical apex a hundred years ago, the mines in

Silver City were minting millionaires, Geronimo was beating the U.S.
Army from his hideout in the nearby mountains, and Billy the Kid was
shooting his first man down on Main Street.

Born in 1952, James Doyne Farmer, who goes by the second of his
two given names, which is pronounced *Dó-an*, is the elder son of a
mining engineer father and entrepreneurial mother. James Doyne
Farmer Sr., who goes by the *first* of his two given names, worked in
copper and iron mines, first in New Mexico, and then for seven years
in Peru and Venezuela. Later, when Jim and Evelyn Farmer redi-
rected their skills toward culinary engineering, they opened a commercial pie shop in Fort Smith, Arkansas.

Born in 1954, Norman Harry Packard is the eldest of six children.
His parents were schoolteachers, with his mother handling second
grade and his father teaching junior high school mathematics. Linzee
and Mary Packard lived in the biggest house in Silver City, a rambling
twenty-three-room brick residence, which was large enough to house
various boarders, Doyne among them. He lived with the Packards for
several months after a military coup in South America frightened his
parents into sending him back to the States. The building also housed
Norman's first two attempts at going into business. Downstairs, facing
the post office, was the Silver Aquarium, a tropical fish store whose
hours were "after school." Upstairs was Norman's TV repair shop,
where he did contract work for Colby's Appliance at the rate of one
dollar fifty cents an hour.

That Silver City produced not one but two world-class physicists is
due in part to the anomalous presence in town of Tom Ingerson. A
stocky man with a domed forehead and piercing blue eyes, Ingerson
had a freshly minted Ph.D. in Einsteinian cosmology from the University of Colorado when he found himself employed as the sole
member of the physics department at Western New Mexico University. This is the old Territorial Normal School, or teachers college, in
Silver City. Ingerson had made a political mistake. One of his dissertation advisers was Frank Oppenheimer, brother of Robert, who led
the team that built the atomic bomb at Los Alamos. Both Oppen-
heimers were later accused of being fellow travelers, and no one
in the Sputnik-crazed 1960s, save for the desperate WNMU, was

willing to hire "Communists" or anyone even remotely associated with "Communists."

To keep the monster of boredom at bay during his four years in Silver City, Ingerson opened his house to the boys in the neighborhood as Explorer Post 114. The kids rebuilt motorcycle engines in Ingerson's kitchen sink. They soldered radios and assembled rockets in his living room. Later they turned his yard into an auto shop devoted to transforming a Dodge van, called the Blue Bus, into a mobile crash pad that carried Ingerson and his extended family on trips all the way from Alaska to Panama.

For Doyne and Norman, Ingerson was the embodiment of ratiocination, and he was the force who directed their natural scientific inclinations to physics. Trained as a theoretical astronomer, he was a gifted tinkerer who knew his way around computers, electronics, telescopy, and mechanics. He was also a dreamer, and one of his romantic schemes involved developing an inherited claim to a sixteenth-century Spanish mine, which was reputed to hold nineteen tons of gold. "Money is the key to freedom," Ingerson told his explorers. "There are two ways to make it, capitalism and theft." Theft is too risky. So that leaves capitalism. Ingerson and the boys filled notebooks with potential inventions and talked endlessly about ways to turn their ideas into gold. Eudaemonic Enterprises was founded under the sign of Ingerson. So, too, one suspects, is the yet-to-be-named company for beating the world financial markets.

Doyne followed Ingerson north to live with him in Moscow, Idaho, when he began teaching at the University of Idaho. Here Doyne simultaneously finished his last year of high school and his first year of college, where his introductory physics course was taught by Ingerson. Then it was time for Doyne to strike out on his own. He headed for California in 1970, where he finished college at Stanford, with a major in physics and a minor in sex, drugs, and rock and roll. In the meantime, Norman Packard had also lit out for the coast, heading for Reed College in Portland, Oregon. The two Silver City sidekicks found themselves back together again when they started graduate school at the University of California at Santa Cruz.

Norman and Doyne are intellectual peers who have developed a remarkable sympathy for each other. They bounce observations back and forth in rapid-fire volleys. They finish each other's sentences, charging forward into new terrain at the mere hint of some half-formed idea. They wrote scientific papers together at Santa Cruz. They collaborated on some of the early, key discoveries in chaos theory. When they split up to pursue their professional careers—Doyne at Los Alamos National Laboratory, Norman at the Institute for Advanced Studies in Princeton, New Jersey—they still managed to meet several times a year and carve out large chunks of time to work together on research in chaos theory, nonlinear dynamics, self-organization, artificial life, and other intellectual passions. No wonder, when given a second chance at starting a company together, Norman opts for Griffin Street over Wall Street.

The equity debate—with some bluster and posturing in the midst of an incredibly frank discussion of how to start a company and value people's work—continues for the next few days. Doyne finds the negotiations so stressful that he accidentally leaves his car engine running all day. McGill wants shares divided between senior and junior people at a ratio of four-to-one. The junior partners want two-to-one. Pelkey says the ratio should go no lower than three-to-one. They eventually settle on three-to-one, with Doyne, Norman, and Jim McGill getting three shares of stock in the company for every one share given to Tony, John, Joe, Stephen, and Tom. They also settle on a name. As a tiebreaker between Griffin Street Forecasting and Metis, they choose Prediction Company. The name is both anodyne and presumptuous, but it is also appropriate.

Several days after the company's christening, McGill walks in the front door and pulls out of his briefcase nine pieces of embossed paper. There is a moment of reverential silence as the predictors gather around the conference table to scrutinize their stock certificates. "I don't know what this means," Doyne jokes, staring at the numbers and the company's name etched in fancy print, "but it's making me nervous." McGill pops the certificates back in his briefcase and prepares to return them to the lawyers in San Francisco.

The one thing Prediction Company cannot agree on is a company logo. Sketched on the whiteboard in the living room are various candidates. There are lunar and solar eclipses, asteroids, meteor showers, and other planetary conjunctions. There is a dog scratching a flea, a soothsayer peering into a crystal ball, and a picture of Einstein shouting "Eureka!" There is a head with a lightbulb shining on top of it, an ice-cream cone preserved in a bell jar, and Michelangelo's divine hand, with lightning bolts coming out of the index finger.

Prediction Company is officially incorporated on September 12, 1991, too late for that year's solar eclipse, but just in time for Fiesta, which Santa Fe celebrates on the first weekend after Labor Day. Word goes out to friends in Los Alamos and Santa Fe that Prediction Company is hosting an inaugural picnic, which will coincide with the start of Fiesta. The three-day celebration begins with the *entrada* and moves on to the burning of Zozobra. Also known as Old Man Gloom, Zozobra is a forty-foot puppet erected on a hill overlooking the baseball diamond in Old Fort Marcy Park. After an elaborate ceremony, involving costumed dancers and amplified groaning, the monster is torched and eventually explodes in pyrotechnic frenzy as everyone in town shouts, "Burn him! Burn him!"

Zozobra is a latter-day addition to what began, in 1712, as a sober reenactment of de Vargas's Indian conquest. Faithfully, for more than two hundred years, the locals had dressed up as knights and followed the richly robed Conquistadora through her *entrada*. But this act of Christian soldiery had become "dull and commercialized," said Will Shuster, when he thought to liven up the event with a burning monster. Shuster is one of the tuberculosis-inflicted artists, writers, heiresses, anthropologists, and spiritual seekers who began drifting into Santa Fe in the 1920s, where they drank steadily through Prohibition, swapped spouses, and experimented with going native.

Shuster borrowed the idea for Zozobra from the Judas figure who every year is paraded through town and burned "alive" during the Holy Week ceremonies in San Miguel d'Allende. To this Mexican import, with its mix of Christian and pagan symbology, Shuster added strong doses of Indian myth and American kitsch to produce one of the city's great cultural events. He began burning a small Zozobra for friends who gathered in his yard. By 1926, when the first public burning was held in front of city hall, the monster had grown to twenty feet. It soon doubled in size and the event was moved to Old Fort Marcy Park, a WPA outdoor theater conceived specially for spectacle. Shuster orchestrated the ceremony for forty years, and it continues today exactly as he designed it. Zozobra means *anguish*, *despair*, or *gloom*, and each year his send-off relieves the city of its anxiety and ushers in the merrymaking of Fiesta. "You know what Zozobra really commemorates and why we're all so happy when he burns?" asks Doyne. "The end of the tourist season."

On a balmy evening with a deep blue sky draped behind the setting sun, Doyne, Letty, Norman, Grazia, and the rest of Prediction Company, accompanied by a long line of friends freighted with backpacks, blankets, and children, make the short walk from Griffin Street to Old Fort Marcy Park. They pass through a gate and over a bridge before coming to the hillside where Zozobra is tethered on a cross of guy wires and Roman candles. They sit on the grass among picnickers dancing to a mariachi band, as everyone waits for the sun to go down.

In front of them, above a flight of steps designed to resemble the approach to a Mayan temple, looms a forty-foot puppet with blubbery pink lips, triangular ears, black pizza-tin eyes, putrid green cheeks, and a mop of frizzy orange hair hanging over a beetled brow. Zozobra wears a white muslin dress, with black cummerbund, buttons, and bow tie. His arms are akimbo, his fingers outstretched in mocking supplication. As darkness settles over the field and the stadium lights go up, the crowd begins to shout, "Burn him!"

Suddenly the lights are extinguished. The blackness is filled with the sound of conga drums and crashing symbols. Zozobra emits a low moan and waves his arms. He groans. He hurls maledictions on the crowd. He grinds his jaw and glowers. From over the hills dance a procession of white-sheathed Glooms carrying torches. The little ghosts do a turn around the monster, taunting him. The malevolent beat of the drums quickens as piles of tumbleweed are ignited at Gargantua's feet. The Glooms disappear into the night.

Next to dance over the hill are three magnificent butterflies. Then the Fire Dancer, dressed in red plumes from head to foot, leaps onto the stairs at Zozobra's feet. He twirls and gyrates in gleeful approach to the doomed giant. Zozobra begins to flail his arms and moan in agony. From his bulbous lips come howls of protest. The chant grows louder, "Burn him! Burn him!"

Suddenly Zozobra is surrounded by bare-armed men holding aloft burning torches. The Fire Dancer leaps onstage. He does a victory dance around Old Man Gloom and then prances and twirls back down the stairs. He seizes two burning torches and waves them over his head. The conga drums beat faster. The crowd shouts lustily. The Fire Dancer makes his second, final approach to the roaring Zozobra.

The hillside explodes in a blaze of fireworks. The Fire Dancer teases his way toward Zozobra's white skirt. Again he circles the monster, waving his burning torches, before plunging them into Zozobra's body. Red flames lick their way up his skirt. They consume his flailing arms and emerge in a great burst of fire from his still-wailing mouth. Fireworks burst from Zozobra's head. Green flares shoot from his eyes, and red flares from his mouth. The stage dissolves in a fiery fountain of Roman candles, topped by a burning crown of Catherine wheels.

Zozobra is Judas. He is Christ on the Cross, a Kachina, an Aztec god destroyed and reborn every year, an auto-da-fé descended from the Spanish Inquisition, a Mardi Gras mummer, a victim of Western justice, a Klansman's bad dream. One year he was Hitler. Another year he was Mussolini. But whoever he is, he is gloriously dead, and now everyone streams toward the plaza to dance away the opening night of Fiesta.

The Fifty-Million-Dollar Suit

> The market behaves as if it were a fellow named Mr. Market, a man with incurable emotional problems. At times he feels euphoric and can see only the favorable factors, while at other times he is depressed and can see nothing but trouble ahead for both the business and the world.
>
> —WARREN BUFFET

> The markets are always wrong.
>
> —GEORGE SOROS

Down at the Science Hut after Fiesta, everyone is looking for the lead pipe cinch. They pull all-nighters crunching numbers. They produce a flurry of research reports. They coach their learning algorithms through endless iterations. They think nonstop about money, how it flows around the world in structured patterns, and where to get some. In fact, the microeconomic dimension of the problem is becoming crucial. Here is a large group of people with an office and overhead and salaries to be paid out of an income of zero dollars. Another sum of money has to be raised for the company's betting pool. Fifty million dollars sounds like a nice number. So Doyne is sent out to get it. Norman will join him later, but for the moment he is off in Illinois finishing his last semester of teaching at the University.

Doyne has no experience in looking for fifty million dollars, but Letty, who has money in her family and knows these things, suggests he begin by buying a suit. Everyone on Griffin Street agrees. In fact, they even debate whether Doyne's suit should be a company expense. Norman and Doyne routinely share clothing, and other people in the company look as if they, too, will be able to step into the "company suit." Doyne meets Letty at her law office, and they walk downtown to Robert R. Bailey, which is the Santa Fe equivalent to Brooks Brothers. Doyne is toying with the idea of a tweed sports coat and slacks, maybe a western look, with a bolo tie, but the salesperson is adamant. *If you're going to Wall Street, you have to have a suit.*

They end up choosing an Italian wool, blue with a hint of olive green. It is an eight hundred dollar suit, on sale for five hundred dollars. The clerk insists they buy a new pair of Bass shoes and a couple of shirts. Doyne draws the line at a rep tie. He will stick with his old M. C. Escher necktie decorated with purple lizards transmogrifying into birds. Back on Griffin Street Doyne gives a fashion show. Everyone thinks the suit looks like fifty million bucks.

Other than founding a company whose sole product is toe-operated computers dedicated to beating roulette, Doyne and Norman have little experience in business. They worked as paperboys in Silver City delivering the *El Paso Times*. Doyne smuggled motorcycles across the Mexican border, mined iron ore in the Venezuelan jungle, and played poker at the card tables in Missoula, Montana. Norman sold tropical fish and repaired TV sets, before he, too, graduated to working as a Nevada card sharp.

Before he moved to Los Alamos Doyne had never had a checking account or credit cards. He worked exclusively in a cash economy and had never made a financial investment in his life, save for buying a house with his wife. In fact, Doyne's credit rating is worse than nonexistent. It is bad, owing to a college loan that went temporarily unpaid during one of his too-frequent changes of address. Norman's financial history is no better. He owns no house. He has never bought a new car, and he, too, has a bad credit rating from an accidentally unpaid cable TV bill.

Despite this wide-ranging experience, when it becomes known that Doyne and Norman are going into business, they begin getting phone calls from America's biggest banks and investment companies, and a lot of high-net-worth individuals start flying their Lear jets into Santa Fe. Clearly, these people are not coming to discuss motorcycle smuggling and TV maintenance. They want to meet two founders of chaos theory and its nonlinear descendants. They are looking for brainy guys who are rumored to know some tricks for beating the world financial markets.

"I believe that scientists should be artists and rebels," says physicist Freeman Dyson, who was a colleague of Norman's when they worked together at the Institute for Advanced Studies. Doyne and Norman share Dyson's opinion. They are socially progressive members of a generation that came of age resisting the Vietnam War. They embrace everything in life that is the opposite of going into business. Doyne, in particular, has a biblical distrust for rich people. Nonetheless, he has met a lot of high-net-worth individuals during his years in Santa Fe. They are attracted to the city by its art scene and its opera. But it was the Santa Fe Institute that really plugged Doyne and Norman into America's business elite.

SFI was founded in 1984 by the senior fellows at Los Alamos, led by George Cowan, a chemist and former director of research at the lab, who managed in his spare time to make a fortune running the local bank. Cowan believed the world needed an institution devoted to studying complex systems. This includes everything from chaos theory to the origins of life. Cowan also wanted SFI to include economics, because he thought economics set "the boundary condition" for difficult subjects in complex systems. To this end, Cowan organized three landmark conferences on economics. It was the last of these conferences, which mingled physicists with Wall Street traders, that was the proximate cause for the founding of Prediction Company. At the conference, so many people were convinced that Packard's and Farmer's ideas would make money that the two men decided to quit their jobs and go into business themselves.

The first Santa Fe Institute conference on economics, entitled "International Finance as a Complex System," was held in August 1986. A dozen people, including John Reed, president and chief executive officer of Citicorp, and Robert Adams, president of the Smithsonian Institution in Washington, gathered for two days at Rancho Encantado, a dude ranch ten miles north of Santa Fe. Reed flew into town in his company's Gulfstream jet and kicked off the conference with a plea for help. He was up to his eyeballs in Third World debt. Citicorp was facing defaults on bank loans to Mexico, Argentina, Brazil, Venezuela, and other countries, which had soaked up loans until the 1982 recession knocked them into bankruptcy. Reed had lost a billion dollars in one year. He was sitting on thirteen billion dollars in shaky loans, and Citicorp was not alone in this financial bath. The total bill for the Third World debt crisis was likely to be three hundred billion dollars.

What worried Reed, a sober man in his mid-forties, with degrees in metallurgy and business from MIT, was that no one saw what was coming, no one knew how to explain it, and no one knew what to do about it. Professional economists seemed to be off in Cloudcuckooland. They had built an elaborate, highly mathematized theory of the world economy that bore little relation to reality. Neoclassical economics assumes that world markets are in equilibrium. Supply counterbalances demand in markets that are always verging toward perfection. But instead of perfection, what Reed saw were financial shocks and turbulence so severe that they threatened to put him out of business.

To illustrate his point, Reed asked his assistant, Eugenia Singer, to describe the global economic models Citicorp uses to run its business. Singer began projecting overhead transparencies outlining various neoclassical models. These included the Federal Reserve Multi-Country Model, the World Bank Global Development Model, the Whalley Trade Model, the Global Optimization Model, and the Organization for Economic Cooperation and Development's Project Link, which is the mother of all models, with four thousand five hundred equations and six thousand variables. It was obvious to everyone in the room that these models, for all their equations and variables, assumed what they should have been predicting—things such as interest rates and financial volatility—and they were missing the social and political factors that move markets, the crashes, feedback loops, and other dynamic elements that were about to lift thirteen billion dollars out of Reed's pocket.

The scientists in the room were surprised that economic theory, as described by Reed and Singer, is so far from reality. It takes as given what should be calculated. Incapable of predicting the future, it discounts it. It lacks all the information about complex systems in motion that physics and related disciplines have been grappling with for years. Without these dynamic elements, economics will never become predictive. It won't even become *post*dictive.

Next it was the scientists' turn to speak. First up at the podium was Larry Smarr, director of the supercomputer facility at the University of Illinois. Smarr discussed how computers are getting very good at making movies of large complex systems in motion. These movies reveal dynamic structure in ways that the human eye and brain alone are incapable of grasping.

Then John Holland, a computer scientist at the University of Michigan, stepped forward to talk about genetic algorithms and other adaptive systems that he—the father of the field—has been developing since the 1960s. Holland described how a gas pipeline network could be directed by a learning algorithm that got smarter with each passing day. A pipeline network, he suggested, is not unlike the global economy in its complexity.

Finally, it was Doyne's turn to speak. He had been invited to the conference as head of the Complex Systems Group at Los Alamos, which, at the time, was one of two research groups devoted to studying complex systems. The other was Norman Packard and Stephen Wolfram's group, located first at the Institute for Advanced Study, and then at the Center for Complex Systems Research at the University of Illinois. Doyne talked briefly about dynamic systems and what physicists know about them. It is obvious to someone trained in chaos theory that financial markets are incapable of being fully described by linear equations, although this is the box into which they have been squeezed. A different kind of mathematics, using nonlinear approaches sensitive to chaos and chaotic attractors, is required if someone wants to understand the economy and its evolution as a complex system.

In linear equations, two plus two equals four. Linear equations describe straight lines, discrete phenomena, and an exceedingly small portion of our everyday experience. In nonlinear equations, the effect is not proportional to the cause. The straw that breaks the camel's back is nonlinear. A small shove can result in a big push. A system evolving in one direction can suddenly veer off in another. Thermometers and bathroom scales are linear. Financial markets yo-yoing between bubbles and crashes are nonlinear. For centuries, scientists reduced the world to linear equations because these were what they could solve. This changed in the late 1970s, with the invention of the personal computer. Its number-crunching skills revolutionized physics by allowing scientists to calculate nonlinear equations. It also allowed them to iterate these calculations indefinitely, which sometimes produces the surprising result that two plus two does not equal four.

Reed tuned into Doyne's speech with single-minded intensity. Then he got visibly excited by what he was hearing. If Doyne was looking for new dynamical systems to analyze, why not tackle the global economy? To help speed him along this track, Reed offered to send him Citicorp's financial database. Doyne was intrigued by the thought of adding Reed's economic data to the research he was already doing on fluid flows and ice ages. This was Doyne's initial contact with the financial world, and he was starting at the top, not just with Citicorp, but with the biggest, hardest, and potentially most lucrative problem of them all.

Everyone at the Rancho Encantado meeting agreed they should follow this introductory discussion with a full-fledged conference on the world economy. This second gathering, called "The Economy as an Evolving Complex System," brought two dozen economists and scientists together in September 1987 for ten days. The Santa Fe Institute was housed by then in the former Cristo Rey convent, an old adobe church tucked among the art galleries lining Canyon Road. The nuns' bedrooms had been converted into offices, and the former chapel became the conference room. In place of the altar was a blackboard, lit with multihued shafts of light from the church's stained-glass windows.

The economics conference was chaired by two Nobel laureates, Philip Anderson, a condensed-matter physicist at Princeton, and Kenneth Arrow, a mathematical economist at Stanford. Anderson had won his Nobel prize for studying phase transitions in which certain metals can flip from being insulators to conductors of electricity. Arrow had won his Nobel for proving that capitalism is the best of all possible worlds. Actually, Arrow only proved that under certain conditions the free market will allocate resources efficiently. Using a lot of fancy mathematics, he and Gerard Debreu had developed this idea in the early 1950s. Called general equilibrium theory, it assumes markets work with perfect information, perfect competition, and absence of technical change. Arrow's work was hailed as a victory for Adam Smith's "invisible hand." It was the intellectual justification for capitalism and proof of free-market efficiency. No one at the time noticed that Arrow had assumed what he should have proved. He described a perfect world, but not this world. The fact that he was willing to debate the issue with a bunch of physicists in New Mexico showed that he knew the question needed revisiting.

Doyne spoke at the conference about new approaches to nonlinear modeling and their application to economic forecasts. In a laboratory experiment with transitional turbulence, he and his graduate student, John "Sid" Sidorowich, had demonstrated that time series analysis, state space embedding, and other nonlinear methods could "give results fifty times better than those of standard linear models." Now someone should see if the new predictive techniques worked this well for economic analysis.

Norman, who was beginning his tenure as a regular visitor to SFI, spoke on the dynamics of development. He had modeled the process with an innovative mechanism that helped explain how new businesses emerge in an economy, while others go bankrupt. It showed how economies function as networks that are continually evolving as new dimensions are added to the economic state space. Norman's model derived from his work with Doyne and biologists Alan Perelson and Stuart Kauffman on self-organizing systems, such as the human immune system, and the origins of life.

"Modeling economic and financial systems is very close to modeling biological systems," he says. "Once I got into it, I found finance even more tantalizing than biology. Where biology takes billions of years to evolve, financial systems recombine and mutate right before your eyes."

Norman concluded in his SFI presentation that the world economy is a complex system, which is evolving according to the same dynamic rules as other complex systems. The economy is a good example of what John Holland calls an adaptive nonlinear network, systems characterized by intensive nonlinear interactions among large numbers of changing agents. Other examples of ANNs are the central nervous system, ecologies, immune systems, the developmental stages of multicelled organisms, and the process of evolutionary genetics. As one can see, this is a very different way of looking at the world than classical economics, which builds its mathematical models out of rational agents who operate in a linear, static, and statistically predictable environment.

When they were published in 1988, the conference proceedings described the event as "a serious dialogue between disparate, and at times hostile, communities of scholars." The physicists thought the economists were bluffing, and the economists thought the physicists were suffering from the Tarzan syndrome. The physicists harried the economists in the question-and-answer periods, and the economists got huffy in defending their discipline. In spite of the heat that was generated with the occasional flashes of light, Doyne and Norman came away from the meeting excited about the problems in economics and convinced they had some fresh ideas on how to solve them.

Neoclassical economics is a highly elaborated, some would say decadent, enterprise that wraps its assumptions in a veil of mathematical equations. Physics is the most math-intensive science, short of mathematics, and the physicists in the chapel had no trouble peering through the equations scrawled up on the Cristo Rey altar to see that the economists' revelations were not divine. The intellectual origins of neoclassical economics are found in the oft-cited, but little-read work *Eléments d'économie politique pure*, which was published by the French economist Léon Walras in 1874, but not translated into English, under the title *Elements of Pure Economics*, until 1954. Walras began the practice of expressing economics in mathematical form.

He analyzed how prices in purely competitive markets, which are governed by the countervailing forces of supply and demand, will fluctuate up and down in the search for equilibrium. The entire Walrasian universe, and most of modern economics until recently, is based on the idea that markets, driven by perfect agents, will eventually fine-tune themselves into static perfection.

The model assumes that competitors are playing a fair game in a world that never sees the kind of technical change that sometimes gives one player a crushing advantage over another. The Walrasian world is endowed with magic marketplaces in which private greed is invariably transmuted into public good. It is a clockwork universe borrowed from Newtonian mechanics. The theory celebrates Adam Smith's claim that the "invisible hand" of free market enterprise will allocate resources efficiently, even in the absence of government intervention. Neoclassical economics is the ideological bedrock of capitalism: a pleasant fiction, but bad science. It ignores what the rest of the world has learned since the *Principia Mathematica* was published in 1687. Even Newton himself, after losing his fortune in the South Sea Bubble of 1720, developed a healthy skepticism about the markets. "I can calculate the motion of heavenly bodies, but not the madness of crowds," he lamented.

The Walrasian theory of economic equilibria has had a difficult time coming to grips with market crashes, bubbles, euphoria, speculation, herd behavior, monopolies, insider trading, and other real-world factors that actually drive the markets away from equilibrium into chaotic cycles and turbulence. Much of the mathematics that dresses up economic theory is borrowed from physics, which explains why economics is full of terms such as *equilibrium, flow, function, vector*, and *external shocks*. But while physics in the twentieth century has witnessed tremendous advances, particularly in the sciences of prediction, neoclassical economics remains frozen in a mechanical world that has yet to catch up to Einstein.

The third big economics conference at the Santa Fe Institute, called "Wall Street and Economic Theory: Prediction and Pattern Recognition," was held in February 1991. Two and a half years after the last big showdown, it was a different sort of affair: far more practical, both for the physicists and for the financiers involved. This time the guests included Wall Street traders and investment bankers from Salomon Brothers, Goldman Sachs, Kidder Peabody, Bear Stearns, and Tudor Investments. Surely, thought the physicists, these market wizards, who daily put billions of dollars into play, will know what makes the global economy tick.

Again, most of the finance people professed to believe some version of general equilibrium theory. What Wall Street does with neoclassical economics is to take the idea of perfectly equilibrated markets—which they obviously are not—and turn the idea on its head, to come up with the notion of perfectly random markets, which they might at first appear to be. According to this theory, one assumes that rational, equally well-informed and competitive agents are betting against one another in efficient markets—*efficient* meaning *random* and *unpredictable* in adjusting to the "correct" price. These assumptions, called the efficient market hypothesis, owe their origin to another Frenchman, Jean Louis Bachelier, who received a "B-" for the idea when it was presented as his doctoral dissertation in 1900.

According to Bachelier, stock market prices look as if the Demon of Chance has drawn them from a bag full of random numbers. The one most-likely bet for where a stock price will be tomorrow is where it was today, and any bet up or down from that price is equally likely, or unlikely, no matter who is making it. The Brownian, or random, motion of stock market prices is described in the literature of finance as a random walk. The definition of a random walk is so lurid that the reader should be reminded of what Oscar Wilde said about the fall of the rupee: "Even these metallic problems have their melodramatic side."

A random walk is a series of sequential moves that resembles the path followed by a drunk lost in a wheat field. He staggers to the left. He staggers to the right, wandering aimlessly through the night. Our only solace in thinking that stock market prices follow a random walk is the presumption that Morgan Stanley, Goldman Sachs, and you and I are equally adrift in this financial wheat field. Future prices are unrelated to past prices, and there is no way, save by closing our eyes and yelling at random, *buy!* or *sell!* that we can hope to make money on the stock market.

Entailed in the efficient market hypothesis is a kind of fatalism. No one can predict future price movements, and all information about these movements is immediately "discounted," or factored into current prices. Therefore, it is impossible to find undervalued stocks, predict the direction of the market, or select winning portfolios by any means other than chance. In several famous experiments that appear to prove their point, random walkers have shown that stock market portfolios selected by throwing darts at the financial pages of the *Wall Street Journal* perform as well as those chosen by professional market analysts.

For the past twenty years, most academic papers in finance have begun with a description of markets as random walks. This enterprise is geared toward making ever-more-elaborate theories proving that markets are efficient. The few academics who have begun to note the ways in which the markets are *in*efficient have a hard time arguing against a theory that now comes in three flavors, strong, mild, and weak, and which has a built-in fudge factor, a category known as "risk," which is a little knob that can be tweaked in any direction required to keep the theory intact.

Evidence countering the efficient market hypothesis comes in the form of stock market anomalies. These are events that violate the assumption that stock returns are randomly distributed. They include the size effect (big-company stocks outperform small-company stocks or vice versa); the January effect (stock returns are abnormally high during the first few days of January); the week-of-the-month effect (the market goes up at the beginning and down at the end of the month); and the hour-of-the-day effect (prices drop during the first hour of trading on Monday and rise on other days). Prices fall faster than they rise; the market suffers from "roundaphobia" (the Dow breaking ten thousand is a big deal); and the market tends to overreact (aggressive buying after good news is followed by nervous selling, no matter what the news). Finally, the efficient market hypothesis is incapable of explaining stock market bubbles and crashes, insider trading, monopolies, and all the other messy stuff that happens outside its perfect models.

Another argument against the random behavior of markets is the existence of speculators, such as Warren Buffet and George Soros, who make winning bets over the long haul. These men can afford to be laconic in their assessment of the efficient market hypothesis. "When the price of a stock can be influenced by a 'herd' on Wall Street with prices set at the margin by the most emotional person, or the most depressed person, it is hard to argue that the market always prices rationally," says Buffet. "In fact, market prices are frequently nonsensical."

Soros is even more emphatic. "The markets are always wrong," he says.

"Buffet and Soros are lucky monkeys," sniff the random walk economists. Their success is a fluke. Give a million monkeys typewriters, and one of them might accidentally type some lines from *Macbeth*. As for the other supposed stock market anomalies, these are nothing more than spurious correlations that exist by chance.

These protests were already sounding strained before the scientists presented their evidence from dynamical systems research, machine learning, time series analysis, and other computer-aided techniques for finding order in chaos. The scientists had yet to spend much time looking at markets, but they were eager to explore the possibility that financial data—like sunspots, ice ages, gas pipelines, and the Italian government—are not random in their behavior, but actually display all sorts of patterns and underlying structure. Many of these patterns are undoubtedly too small to trade on, given the cost of executing these trades, but other patterns might cut through the markets like a rich vein in the mother lode.

On the first day of the conference the Wall Street practitioners had described how they model financial markets. On the second day the scientists had presented their techniques. On the third day the two groups get together to brainstorm how to apply these new techniques to financial markets. Norman and Doyne are excited by the possibilities, and they think it might be fun to get involved with some of the traders, who seem like decent people. But as the day wears on, the economists in the audience begin pushing the discussion back toward efficient markets and the impossibility of predicting them. This provokes Doyne into losing his patience.

"Wait a minute," he says. "During the first day we heard people who actually trade for a living describe how they make large sums of money. Why are we wasting our time on a discussion of whether or not markets are efficient? Let's just assume profits are possible and figure out how to make them."

After the Cristo Rey meeting, Doyne and Norman know what they have to do to prove that dynamical systems research can beat the markets. They have to do it.

A Better Class of People

Of all our human inventions, economic man is by far the dullest.

—GREGORY BATESON

The important thing for every man to learn is that money is not to be despised.

—HENRY MILLER

In early September, two months after the founding of Prediction Company, Doyne phones "Zeus" Pelkey, as their Olympian investor is nicknamed, and tells him he is ready to go on the road. Pelkey phones back in an hour to tell Doyne he has meetings scheduled with the ex-chairman of Bank of America and the director of Montgomery Asset Management, part of Montgomery Securities.

Doyne packs his new suit and flies to San Francisco. He will spend the night at Jim McGill's apartment on Nob Hill and give his first sales pitch to Bank of America in the morning. McGill asks Doyne to show him his presentation. He takes one look at the handwritten transparencies and shakes his head. "We have to redo them," he announces sternly. "This kind of thing is all right for an academic audience, but not for the chairman of Bank of America."

They go to work on McGill's dining room table, transferring Doyne's data to computer-generated images. By three in the morning they have a snappy presentation with razor-sharp graphs printed on a corporate-blue background.

For the predictors back on Griffin Street, McGill is Mr. Business. He incarnates a kind of no-nonsense, profit *über alles* mentality. He seems happier running spreadsheets than talking to employees, and he seems happier having employees than friends. He prefers hierarchy to democracy, and he doles out information to underlings on a need-to-know basis, which is seldom. His driving passions seem to be management software, financial controls, and spit-and-polish quarterly reports, which are more appropriate for a Fortune 500 company than the chaotic flotsam gathered in Santa Fe.

Under his tough-guy stance, McGill is actually an old California surfer who likes fast cars, good food, jazz, Hawaiian beaches, and Asian women. His girlfriend is Thai. He keeps a Columbian fertility fetish on his desk. He drives a BMW with a machine-tooled dash that looks as if one of its switches is for firing retrorockets. He studies kung fu from a master in San Francisco and delves into Eastern mysticism. Doyne describes him as a "Zen Buddhist businessman" who carefully controls how he presents himself in the world. He believes finance is a martial art that requires skill in psychological and other kinds of combat. McGill gives Doyne two precepts to follow as they go in search of a partner: *Never say anything bad about your former employer*, and *Always wait for the other person to make a move before you respond*.

Because he keeps his personal and business lives divorced, McGill is a cipher on Griffin Street. He commutes to Santa Fe from San Francisco, occasionally dropping out of the sky to install new computer software or remind people to file their quarterly objectives. He spends hours on the phone and more hours rushing through airports on his way to meet all the big bankers in America who are toying with the idea of investing in Prediction Company. He is not building market models or doing anything else that appears productive; so people on Griffin Street wonder why they need him. But Doyne and Norman know all too well why they need him.

"Prediction Company is the opposite of Eudaemonic Enterprises," says Doyne, referring to their old roulette company. Eudaemonic Enterprises was a hang-loose, radically egalitarian project run out of a California commune on donated labor and a dream. It was great fun, and it failed. At least it failed to make a lot of money, which is why Doyne and Norman think that this time they should try a different kind of corporate structure. Doyne summarizes the mistakes he plans not to repeat. "Don't work on a shoestring; don't try to do everything yourself, like building computers from scratch; and avoid the temptation to hire troublesome geniuses."

Born in 1945, the son of an air force pilot and stepson of an electronics engineer who managed a plant that made high-voltage power switches, McGill grew up in Monterey, California. "I was the hardworking, high-grades, competitive type," he says, before admitting that he also spent a lot of time surfing off Asilomar Beach and rebuilding car engines. "If I hadn't bought so many cars, I'd be rich by now."

McGill went to college at the University of California at Santa Barbara. In 1967 he moved to graduate school at the University of California at San Diego, where he made what he describes as "one of the biggest mistakes of my career." Linus Pauling, two-time Nobel laureate, asked McGill to spend the summer with him on his ranch in Big Sur, helping to revise his book on physical chemistry. Wanting to focus on his dissertation, McGill turned him down.

Tired of being teargassed on a campus that was in the throes of daily demonstrations against the Vietnam War, McGill moved to the newly opened University of California at Santa Cruz, which was being built in a redwood forest, on a stunningly beautiful site that looked south over Monterey Bay toward McGill's old home on Point Lobos. McGill was among the first graduate students to enroll. Another earlycomer was Rob Shaw, with whom McGill shared adjoining offices. Both of them were working on experiments in low-temperature physics until, a few weeks shy of finishing his Ph.D., Shaw threw away years of research to start dabbling in a field so new that only later would it be called chaos theory. Shaw was the Pied Piper who led Doyne and Norman, along with Jim Crutchfield, into forming the Chaos Cabal.

McGill barreled on. He finished his dissertation on ripplons, which are acoustic waves produced by squirting electrons onto liquid helium, and graduated from Santa Cruz in 1974. He had two job offers. One from the University of Hawaii and one from a start-up company in Santa Barbara, California. He chose the latter.

"I never thought I wanted to go into business," he says. "My stepfather was in business, and for a kid with a healthy Freudian urge to kill his father, it was last thing I thought I wanted to do. But when it came down to it . . ."

McGill was hired by a company called Culler Harrison to research voice signal processing for the Arpanet, the military precursor to the Internet. His assignment—twenty years before it became standard practice—was to get telephone conversations transmitted over the nation's first computer network. McGill and his colleagues built a box that converted human speech into a string of digits and then compressed it into a bit-per-second package that was fifty times smaller than what anyone had previously done. It was a nice piece of work, and something like it is now incorporated into every telephone system in the world.

McGill branched out from working on speech signals to music. He landed a contract from CBS to develop new keyboard instruments. These were the first instruments to employ the novel technique of storing sound in computer memory and playing it back at full fidelity. Music has a much higher information rate than speech, and computer chips were so expensive back in the 1970s that the trick to making these instruments work was the same kind of data compression that Culler Harrison had developed for the Arpanet. In 1978 McGill started his own company to carry the idea forward. It began as a two-person consulting shop called Digital Sound Corporation.

McGill was president. He didn't get a paycheck for two years. The company eventually found a niche for itself building high-performance IO processors, which are machines for moving large amounts of data from one place to another. In this case, McGill's processors were moving high-quality sound from microphones onto computer discs. Then in 1981, McGill had one of those entrepreneurial flashes, when a bright idea meets a commercial opportunity and a business is born. He was consulting at Wang Computers in Massachusetts, when he noticed lying on someone's desk the prototype for one of the world's first voice-mail systems. The engineer obliged McGill with a demo, and he saw right away that what the system lacked was voice compression. McGill rushed back to California to write up a business plan. He secured two and a half million dollars in venture capital and started building voice-mail machines. This was the second incarnation of Digital Sound Corporation.

Soon McGill had raised another fifteen million dollars in capital and was selling eight million dollars a year of voice-mail equipment from a shop with one hundred and fifty employees. Anyone with a head for business will know that this is a company in trouble. The sales figures are too low to support the debt and production costs. "I was in over my head," McGill admits. His company got taken away from him by his bankers, who installed a new president. Within a year, McGill had sold his stock and moved on to another project.

In 1987 he began working at Telebit, which at the time was a small Silicon Valley manufacturer of computer modems. McGill was hired to run Telebit's engineering and marketing departments, but he found himself getting increasingly interested in the business side of business. "Commerce is deeply wired into civilization, and everyone sells something," he says. A decade after he arrived at Telebit, the company was sold to Cisco Systems for two hundred million dollars, but long before then McGill was off on another adventure.

In 1991, when Telebit had grown into a company with four hundred employees and sales of fifty-five million dollars a year, McGill quit. Again there was a new president he didn't like, and McGill was ambitious. He realized he was good at chasing venture capital and riding the growth curve of early-stage technology companies, but he had yet to make a fortune. "My batting average is so-so," he says. "I'd like to have a big hit."

McGill had been working seventy hours a week and living on airplanes, commuting to Geneva, where he engaged in marathon battles fighting UN bureaucrats over datacom standards. So when he left Telebit, the first thing he had in mind was a long vacation. He was planning to fly to Asia for four months, where he would do some body surfing in Bali and hiking in Thailand. But before he got on the plane, he received a call from Doyne, who was talking about starting a company to predict the future.

After staying up until 3:00 in the morning to whip his presentation into shape, Doyne and McGill, and Jim Pelkey, who has flown into town for the meeting, arrive at the office of Lee Prussia, emeritus chairman of the board at Bank of America. Prussia is impressed, but nonlinear dynamics is Greek to him. So he gets on the phone and arranges two more meetings. "What happens after talking to the big guy," says Doyne, "is that he sends you downstairs to talk to someone who can evaluate what you're saying, and then this person reports back to the head guy."

A few days later, Doyne is repeating his presentation for Terry Turner, a former professor of finance at UC Berkeley and treasurer at the bank. Turner's eyebrows shoot up when he sees Doyne's graphs, which show Prediction Company getting a ten-percent unleveraged return on its investment. If the investment were leveraged—using futures contracts, options, and other financial bets made with borrowed money—the return could go up twenty-fivefold. Prediction Company is forecasting whopping returns, but it is also forecasting whopping exposure to risk. "These are smart guys," he reports back to Prussia, "but they have to hunker down for three months and work like the devil to get their ideas pulled together."

Prussia's second phone call is to one of Bank of America's proprietary traders. Prussia thinks it might be useful for Prediction Company to shadow the guy for a day and compare notes. This time it is Norman's turn to wear the company suit. He flies to San Francisco and reports to the bank to watch one of America's largest financial institutions play the world markets. He is surprised when he is introduced to a man and his assistant ensconced in a small room, whose walls are covered with sawtooth graphs. The trader, using a ruler and compass, has drawn various lines on these charts, which he is using to predict whether the markets will rise or fall.

Norman has stumbled into the lair of a chartist, an occult tape reader who thinks he can predict market moves by eyeballing the shape that stock prices take when plotted on a piece of graph paper. Chartists are to finance what astrology is to space science. It is a mystical practice akin to reading the entrails of animals. But its newspaper of record is *The Wall Street Journal*, and almost every major financial institution in the United States keeps at least one or two chartists working behind closed doors.

The trader and his assistant sit at two desks, each of them facing a wall of green cathode ray tubes. Scrolling over the screens are news feeds and price data. Installed below the screens is an array of push-button speed-dialers wired directly to brokers on the floor at the San Francisco, Chicago, and New York exchanges. Squawk boxes connected to these floor brokers pipe back into the bank a steady patter of contracts being bought and sold. All day the trader and his assistant watch the numbers, doodle on their charts, and speed-dial orders for currencies, gold, commodities, futures, and options. "Don't believe what you see on the screens," the trader advises Norman. "Go with the voices off the floor. Go with the juice in the market."

The trader, a tassel-loafered gentleman in his fifties, enjoys telling Norman about the good old days when stock market speculating was reserved for guys with iron balls. Norman mentions that a computer, which can look at thousands of variables—as opposed to a human being, who can look at only a few—might be good at scanning market charts and finding telltale patterns. "No, thanks," says the trader. "There's nothing a computer can do that I can't do. I'd rather validate the numbers by intuition."

Norman describes how Prediction Company is hoping someday to develop a news reader, a computer good at analyzing political events and evaluating what traders call *market sentiment*. "I *avoid* looking at the news," says the trader, echoing a central precept of market charting. "The news is in the tape. Everything you need to know is in the numbers. If you stare at the screen and know how to read it, you can see the news before it happens. The Gulf War, for example. You could tell from the price of oil futures that the war was going to happen a month before it started."

Chartists trace their lineage back to Charles Dow, who founded Dow Jones & Company, in 1882, and used it to publish his ideas, later called Dow Theory. Dow believed that the stock market moves in waves. It rises or falls for extended periods. It maintains these regimes, until, at key moments in the life of the market, it reverses direction. Dow thought he could predict these reversals by discerning patterns in the graphs of stock market prices. The Dow Jones Industrial Average, which is cited on the nightly news as the very barometer of American capitalism, began its life as Dow's tool for charting stock market patterns.

To track the market, Dow summed the closing prices of the stocks for a few big companies. After he began publishing his average, Dow noticed that it tended to hesitate when approaching round numbers, the number three thousand, for example. The market would pause, as if facing a hurdle. But once the market had screwed up its courage and "penetrated a resistance area," it would tend to keep on going. The stock market represented for Dow a Manichaean struggle between *resistance* and *support*. When the market approaches a previous high, but fails to rise above it, it is succumbing to resistance. When it approaches a previous low, but fails to sink below it, it is being supported by some kind of residual memory. The exciting moment for Dow and his fellow chartists comes when the market "penetrates a resistance area" or swoons through its supports. These are the moments when Prechter, Granville, Garzarelli, and other famous chartists broadcast orders to their loyal followers—orders that by themselves can move the New York Stock Exchange hundreds of points up or down.

Dow's original insights were later developed by John Magee, the author, along with Robert Edwards, of the chartists' bible, a book called *Technical Analysis of Stock Trends*, which was first published in 1948 and is now in its umpteenth printing. Filled with hundreds of graphs, like those on the walls in the Bank of America trading room, Magee's book illustrates the patterns that chartist's use to predict market moves. In spite of its dry title, Magee's tome is replete with the sexually charged language employed by chartists who are trying to master bull markets. Triple-top patterns, heads-and-shoulders, wedges, diamonds, channels, and pennants develop as markets slide *downward through the neckline*. Markets *penetrate* double bottoms.

They *break through* resistance areas. They *violate the lows*. They *firm up big plays*. They ascend peaks into *upside breakouts*, before finally *attaining a buying climax*. The next time you hear someone talking euphemistically about a stock market *correction*, you will know that you are in the company of a hopeful chartist.

Magee worked out of a small office in Springfield, Massachusetts, where he insisted on reading newspapers two weeks late and kept his windows boarded up, to free himself from outside distractions. The stock market behaves, he said, like a pig in a barn. The pig wears a harness around its body, and to this harness is attached a long flag pole. The barn doors are closed, but the hayloft is open. So all one can see of the pig, as it runs around the barn, is the flag waving on top of the pig pole. Don't "try to 'interpret' these movements as corresponding to various assimilative, combative, copulative, etc., actions of the pig," advised Magee. Don't try to "assign 'meanings' to the pole's movements." Nor should one care if the pig is named Wilbur or Microsoft. To be a successful stock market chartist, just watch the pole!

On leaving their meeting with Lee Prussia, Doyne and the two Jims head to Montgomery Securities, Jim Pelkey's old home, to give another presentation to the director of Montgomery Asset Management. MAM is the wing of the bank that puts money managers together with clients. Their head stock picker has just finished a stellar year. They have money in their pockets. They like Doyne's professional-looking transparencies. They want to do a deal.

Montgomery makes most of its money out of financing Silicon Valley stock offerings. But it also makes money in other areas, such as venture capital and asset management. Asset management is the well-groomed, client-friendly side of the business. It involves spending a lot of time pressing flesh to get rich people and insurance companies to give you their money to manage. This happens during road shows to places like Kansas City and Tulsa. "To clinch the deal we'd have to be Mr. Cool technical guys," Norman explains, "putting fund managers in Iowa in touch with 'chaos' and what it can do for them."

Montgomery promises to limit the road shows to twenty a year. They will bring in Doyne and Norman only for "finals," when the deal gets clinched. They whittle the number down to ten road shows a year. They claim the process is relatively painless and, to prove it, they set up a trial run for Doyne to pitch Prediction Company to David White, who manages the endowment at the Rockefeller Foundation. Doyne gives his presentation. White likes it. He says if Prediction Company signs with Montgomery, he will give them enough money to start trading right away. Doyne's suit is working. But something tells him this isn't the way to go.

After his day in San Francisco pitching Prediction Company to Bank of America and Montgomery Securities, Doyne rises at 5:00 in the morning and flies to Orange County, where he is scheduled to meet Sheen Kassouf and Alan Lewis at a company called Analytic Investments. Analytic, which manages several billion dollars for respectable clients, like Harvard and Yale, is located in one of the prestressed haciendas with smoked-glass windows that line the freeways inland from Newport Beach. This medium density sprawl around John Wayne Airport is the Wall Street of southern California, housing the branch offices, and even some of the main offices, of America's major financial institutions. Analytic's modest quarters are decorated with seascapes and wandering jews maintained by a plant rental service. The place is deadly quiet, and the company is not a suitable partner for Prediction Company. But Doyne is not here for money. He is here for advice.

Sheen Kassouf, a courtly, fastidious gentleman from a family of Lebanese traders, is the person who pioneered the use of the options market by institutional investors. He was a professor of economics at the University of California at Irvine, when he quit to found Analytic Investments in 1970. At the time, pension trusts and endowments were prohibited by the U.S. comptroller of the currency from playing the options markets, which were considered too risky for anyone but seasoned speculators. Forbidden from accepting clients, Kassouf spent three years fighting to change the rules. After winning his case in 1973, he began building a conservative hedge fund, which is now much appreciated by vice presidents of finance at Ivy League colleges and other blue-chip institutions.

Alan Lewis, with degrees in physics from Cal Tech and Berkeley, had worked for a decade at Analytic as their in-house theoretician, before getting kicked upstairs to become president. Lewis, a laid-back westerner who enjoys beach towns and sports cars, is an old roulette-playing friend of Doyne and Norman's. Lewis himself had tried to beat roulette in the early 1970s by building a computer into a camera case. He had played his system in Las Vegas a couple of times before scrapping the project. But he got excited again when he moved to Santa Cruz as a young physics instructor and saw Doyne and Norman's more serious operation. He enlisted as a researcher and joined the team on several forays to Las Vegas.

Lewis is an expert in statistical mechanics, that branch of physics which is good at describing the atomic behavior of gases, and options pricing models. He is not an expert in nonlinear forecasting, which developed after he left academia, but he knows how these ideas have to be packaged if finance people are going to take them seriously. Lewis gives Doyne some pointers on how to polish his presentation and then, over the next few weeks, he keeps pushing Prediction Company to develop a battery of statistics for testing whether their system really works.

Doyne also solicits the advice of another Newport Beach communicant in the holy trinity of gambling, physics, and finance, Edward Thorp. Back in 1960, when this tall, tight-sprung son of a Los Angeles security guard was a newly appointed instructor of mathematics at MIT, Thorp had screwed up the nerve to knock on the office door of Claude Shannon. Shannon, then in his early forties, was the distinguished professor of applied mathematics who had invented information theory. While still a graduate student, he had worked out the equations for comprehending switching electrical networks. He extrapolated outward from telephone exchanges to computers and on from there to the human brain, all the time developing his "basic idea" that "information can be treated very much like a physical quantity such as mass or energy."

Thorp wanted Shannon's help in publishing a scientific paper he had written on card counting. Thorp had developed a system for beating the game of twenty-one or blackjack. The system takes advantage of the fact that a player's probabilities of beating the house change as a deck of cards is dealt down from top to bottom. A deck might become ace-rich, for example, depending on how many aces remain to be dealt. Using MIT's mainframe computer, Thorp had calculated that a card counter with perfect memory and faultless play could gain a three-percent advantage over the house. Thorp in 1962 wrote a best-selling book about card counting, *Beat the Dealer*. Then he conducted a famous demonstration of his system in Las Vegas, with a reporter from *Life* magazine at his elbow.

Shannon appreciated Thorp's research, but what really excited him was another one of Thorp's ideas. This was a Newtonian system for beating roulette. It was based on clocking the game's moving parts and calculating their relative positions. This allowed one to predict where the spinning ball was likely to come to rest. The two scientists went to work in Shannon's basement, analyzing roulette wheels and verifying that the game is, in fact, beatable by physical prediction. They built an analog roulette computer the size of a cigarette pack and launched an attack on the casinos in 1962. After Claude and Betty Shannon and Vivian and Edward Thorp had spent a week at the old Riviera Hotel on the Las Vegas strip, the experiment ended in a snafu of broken wires and electric shocks. It would be another eighteen years before Eudaemonic Enterprises, this time using digital computers, completed what Thorp and Shannon had begun.

Thorp calculated he could make three hundred thousand dollars a year as a card counter. He could make "in the low millions" beating all the roulette tables around the world. But he soon realized there was another gambling game that was *much* bigger: the stock market. After he began teaching at the new University of California campus at Irvine, which was being built in the hills behind Newport Beach, Thorp in the mid-1960s began working on his next winning system. This time he teamed up with Sheen Kassouf, who had just moved to California from New York.

Kassouf, who had recently finished his doctorate at Columbia

University, before taking a job in the economics department at UC Irvine, was looking for a stock market system that would allow him to win, whether the market went up or down. He found what he was looking for when he hit on the idea of hedging warrants. These convertible securities, which are traded like options, are sometimes "mispriced." Kassouf's idea was to buy underpriced warrants in one market and hedge the bet by selling overpriced stock in another. He would make money when the two markets swung back into synch.

Thorp trained his mathematical skills on Kassouf's idea, and together they developed a stock market system for hedging warrants. The system is described in Thorp and Kassouf's *Beat the Market: A Scientific Stock Market System*, published in 1967. This turned into another best-seller for Thorp, and soon the two authors had quit their teaching jobs to go into business, although not the same business. Kassouf took the high road into fund management. Thorp took the riskier road into junk bonds and other speculative investments. In 1969, Thorp and a new partner, James Regan, founded a hedge fund called Princeton/Newport Partners, which they managed from bicoastal offices in Princeton, New Jersey, and Newport Beach, California. P/NP soon grew into a three-hundred-million-dollar company. It specialized in computerized trades of Eurodollar convertibles, junk bonds, and other exotic financial instruments. Jay Regan bought a two-hundred-and-twenty-five-acre horse farm in New Jersey. Edward Thorp built the largest house in Newport Beach, a sprawling hilltop mansion with ten bathrooms and a bomb shelter in the basement designed to withstand a one-megaton blast. Thorp was getting so good at predicting the future that he began thinking about altering it. Overcoming death would be a good place to start. He arranged to have himself cryonically suspended. After he dies, his body will be frozen and then revived at some suitable time in the future.

Jay Regan also appeared to be altering the future. This came to light in 1989, when fifty federal marshals, wearing bulletproof vests, raided his Princeton office, cleaned out the firm's paperwork, and charged Regan with racketeering, conspiracy, and insider trading. One of Regan's good friends and business partners was Michael Milken, the king of junk bonds, who himself was about to run afoul of the law. Regan was accused of parking stock for Milken, which involved temporarily moving assets from one balance sheet to another, in order to create phony tax losses.

This white-collar crime was unusual only because Princeton/ Newport had run afoul of Rudolph Giuliani. Then U.S. Attorney in New York, Giuliani was hoping to boost his political ambitions by mounting a crusade to clean up Wall Street. His weapon in this crusade was the RICO criminal racketeering statutes, which until then had been reserved for catching mobsters. RICO is a legal bazooka good at blowing away the enemy without the usual evidentiary requirements found in civil trials. Princeton/Newport was the first securities firm ever to be labeled "a racketeering enterprise." While Jay Regan was facing a fine of twenty-two million dollars and a twenty-year jail sentence on each of sixty-four counts, for a total of one thousand two hundred and eighty years in the slammer, Rudy Giuliani was elected mayor of New York City.

Thorp was never implicated in this affair, and the charges against Regan were latter dismissed, but not before Princeton/Newport was put out of business and an embittered Thorp had started using a picture of Giuliani as a dartboard. "Wall Street is like a big gambling casino," he remarks. "The game is much bigger and much more interesting to me than casino gambling. I also used to think I was dealing with a better class of people. But I was young and naïve in those days."

Thorp founded a second hedge fund in the late 1980s, Edward O. Thorp & Associates, with two hundred million dollars under management. When Doyne phones him to ask if Thorp wants to invest in Prediction Company, Thorp says, "I would have been interested, if you had phoned me a few months ago, but I just found something that works." With Rudy Giuliani's dart-filled face looking over his shoulder, Ed Thorp is on another roll.

The Badlands of Capitalism

It's all very, very noisy out there. Very hard to hear the
tune. Like a piano in the next room, it's playing your song,
but unfortunately it's out of whack, some of the strings are
missing, and the pianist is tone deaf and drunk—I mean,
the *noise! Impossible!*

—TOM STOPPARD

Chaos is the score upon which reality is written.

—HENRY MILLER

At the end of September, Doyne and Jim McGill fly to New York for
five days. Their destination: Wall Street. Between Zeus Pelkey's
phone calls and Doyne's contacts from the Santa Fe Institute, meetings have been arranged with some of the country's biggest investment banks. Wall Street is the badlands of capitalism, the deep
canyons of high finance, the world's richest vein of speculative gold.
For a kid from Silver City, *this* is the place to look for fifty million
dollars.

Doyne has a wide-open, western informality and the nerve of a
gambler, but he is no hick. He can *play* at being a hick when it serves
some strategic purpose. Clem the Cowboy, a hay-sucking nitwit from
New Mexico, was the disguise he assumed at the poker tables in Missoula, Montana. But Doyne is also a Stanford graduate married into
an old Boston family. He knows how to shake hands at NATO conferences and chat up the military-industrial bigwigs who used to fund
his research at Los Alamos. So it is not too much of a stretch to find

him lunching in corporate boardrooms overlooking the Manhattan skyline.

Norman is endowed with the same informality and directness, although his exuberance, which in Doyne's case sometimes comes across as rudeness, is tempered by an unfailing politesse. Norman is an accomplished pianist in the classical repertoire from Bach to Chopin. He has lived in France and Italy. He has mastered the art of afternoon tea ceremonies in Princeton. So Norman, too, is a credible emissary when he gets to wear the company suit. But Norman is still teaching, during his final semester at the University of Illinois. So these early forays onto Wall Street are conducted by Doyne, usually with Jim McGill at his elbow. McGill helps steer him around town and balance the team with fiduciary gravitas.

Doyne crashes on a friend's couch. McGill installs himself in a midtown hotel. The following morning, as they begin trooping through executive suites, Doyne begins to develop an anthropology of corporate New York from the seventieth floor. "I tune into the layout of these places, which tells you a lot about how they're run. Most of them have glass walls with incredible vistas over the Statue of Liberty or Wall Street or all the way uptown to Central Park. Sometimes there is an open floor-plan, and everyone, even the secretaries, gets the view. Other times, there are lots of walls, where only the bosses get the view."

An old sixties radical, who mistrusts both the military-industrial complex and the fat cats on Wall Street, Doyne is wary about approaching this alien world. "I expected to find the stereotypical greedy bastards, but I ended up feeling positive about most of these people, which really surprised me. I got the feeling they wouldn't cheat me. I knew they'd drive a hard business deal but, for the most part, they struck me as intelligent people with a high degree of integrity."

Doyne's first New York sales pitch is at Kidder Peabody, at 60 Broad Street, between Wall and Water. This investment bank in the heart of the financial district was owned at the time by General Electric, which makes more money from playing with money than it does from manufacturing toasters and television sets. Doyne is surprised to find himself ushered into a conference room with thirty executives expecting him to deliver a lecture. He grimaces on being given a rousing introduction as the man who beat roulette and invented chaos theory and who is now going to beat the stock market.

The traders and executives in the room are wearing broad-shouldered suits and rep ties. Doyne cuts a leaner profile in his Italian suit and Escher necktie. He is all springy intensity, with a Brillo pad of brown hair curling over his ears, and deep blue eyes into which pool a steady flow of ideas. Combining the hard edge of a scientific skeptic with the enthusiasm of a gold bug, Doyne launches into a lecture on financial forecasting. It is a bravura performance.

"Everyone knows that prices in financial markets move up and down, but no one knows why or how," he begins. "Economists claim these price moves are a random walk. They are the unpredictable product of efficient markets. Prices in these markets reflect the activity of rational, logical, and always equally well-informed investors. I don't know about you," he comments, "but I'm not always a rational human being, and I think this is a pretty far-fetched view of the world."

Many Wall Street traders scoff at the idea of efficient markets. In their line of work they often see what look like patterns in the oscillating prices of stocks and bonds. But the big question for Doyne, and for everyone else in the room, is whether a method exists for finding these patterns. So far the tools have been inadequate to the task. Conventional forecasting technology, which matured in the 1960s, is linear. But financial time series reveal little or no linear structure. So the technology required for this assignment has to be imported from elsewhere, and the best place to look is the new science of nonlinear prediction.

"At Prediction Company we assume the random-walk hypothesis is wrong," he declares. "There are patterns in market data, they appear more often than one would normally expect, and they reappear in the future. We use black-box models to extract the predictive value from these patterns. We monitor many inputs, continually evaluating which ones are relevant. We let the data speak for themselves. We look for pockets of predictability, shifting regimes where order can be found emerging from what are otherwise highly chaotic time series."

Black box computer models are data intensive. They require numbers pouring into them from around the world. Models with hundreds of different data streams, even when run on supercomputers, like those at Los Alamos, tend to be clunky and slow. So the trick is to isolate the important inputs and pop them into speedy little nonlinear models. In the best of all possible worlds one would create a voting culture, in which dozens of predictive models continually poll themselves on market sentiment, volatility, price moves, and other key factors that dictate whether the markets will go up or down.

Doyne gives an example of the kinds of models Prediction Company is building. "Let's say the differential interest rate is falling, market momentum is low, and volume is rising, and that when these conditions were met over the past year, currency prices rose nine out of ten times. There are a large number of patterns like this. The trick is to find the good ones."

Doyne spells out for his audience how nonlinear forecasting should not be confused with the charts employed by Charles Dow and John Magee. Nor is Prediction Company using methods based on artificial intelligence. Its technology is borrowed from the physics of complex systems. This is new technology, which has not yet been applied to finance, although it has already proved successful in developing predictive models for fluid turbulence and weather patterns. When applied to market data, the new technology produces substantial returns, and there is less than a one-in-ten-thousand chance that the models are wrong.

As Doyne finds himself giving more speeches to Wall Street financiers, his attack on the efficient-market hypothesis becomes more pointed, and his lecture gets illustrations. Included is a graph of crude oil prices from 1980 onward. This graph in no way resembles a random walk. In fact, the graph begins oscillating wildly in 1990, before the outbreak of the Gulf War. "It is obvious that the news is heavily influencing the financial markets," he notes. "If you want to know what the markets are doing, you have to be able to predict the mass of human behavior."

Graphs showing the probability distribution of random events in nature tend to take the shape of Karl Friedrich Gauss's famous bell-shaped curve, which represents a normal distribution of data points. Most-likely events cluster under the bell. Least-likely events straggle out toward the edge of the graph. There are exceptions to this rule, one interesting example being a graph of Mississippi River flood levels. Although one might expect this graph to be Gaussian, it is not. It possesses what are known as *fat tails*. Clustered along the x axis, where improbable events are measured, is a bulge of data points, which indicates that major floods or droughts occur far more frequently than one would normally expect. The graph is similar to what one finds when looking at financial markets. They, too, have fat tails, indicating a propensity for extreme events.

The most arresting illustration in Doyne's talk is a black-and-white photograph of what looks like the chaos from which God created the world. Night and day swirl around each other in an inchoate mass. But as one stares at the picture, patterns begin to emerge. One sees vortices of night curled into whirlpools of day, and these starry patterns repeat themselves endlessly, like a mirror image reflected back on itself. Doyne keeps a framed copy of this picture on his office wall.

Famous among aeronautical engineers and fluid dynamicists, the picture first appeared in a scientific paper, "On Density Effects and Large Structures in Turbulent Mixing Layers," published in 1974 in the *Journal of Fluid Mechanics*. The two authors of the paper were Garry Brown, a postdoctoral student, and Anatol Roshko, a professor of aeronautical engineering at the California Institute of Technology. Brown would later go on to direct the aeronautics department at Princeton.

Brown and Roshko's photo shows two gasses shot at high speed through a wind tunnel. The gases are tumbling around each other in a display of what aeronautical engineers call turbulent fluid flow. Leonardo da Vinci is the first scientist to describe this phenomenon, and he is the first to draw pictures of *turbolenza*, as he called it. His sketches of vortices and whirlpools in turbulent flow are contained in a notebook later called the *Codex Leicester*. The book is currently owned by Microsoft chairman Bill Gates, who has installed it as the centerpiece of the library in his Seattle mansion.

Turbulence is key to designing airplane wings. It is crucial for mixing gases, predicting the weather, and firing rockets. Every moving object encounters turbulence, and the ultimate objective for applied mathematics is to predict turbulent flow. "Turbulence is the greatest puzzle in classical physics," declared Nobel laureate Richard Feynman.

Ever since Leonardo, scientists have known that patterns of organized structure exist in turbulent fluid flow. This can be observed by anyone lazy enough to sit beside a rock-filled stream on a summer day. The stream forms little whirlpools that come together and break apart as the water gurgles downstream. Yes, at low speeds one can find pockets of order within the chaos of turbulence. But what does one see at high speeds? What happens when the stream is a raging torrent?

For centuries this question provoked a vitriolic debate among scientists. One camp said the structure, the little waves and vortices, still existed in high-speed flows, even if no one could see them. The other camp said this was impossible. The structure disappeared into utter randomness. Nothing remained but the chaos of molecules tumbling around one another in Brownian motion. The debate lasted from the Renaissance until 1970, when Brown and Roshko settled it by capturing their remarkable picture of order in chaos.

They had built a wind tunnel equipped with nozzles for mixing gases flowing at different speeds. When a high-speed stream flows next to a low-speed stream, the interface between the two becomes turbulent. Brown and Roshko's apparatus was outfitted with a high-voltage spark for a light source and a device for taking pictures of the turbulent region. Their famous image shows two gases, helium and nitrogen, flowing next to each other. Where the two gases are mixing, one sees "coherent structure"—swirling white vortices, like waves breaking in the ocean.

At the edge between the two gases, reappearing over and over again, are kaleidoscopic inlays, as neatly patterned as snowflakes. These "coherent" patterns, which nest inside one another and reappear at different scales as the gases flow downstream, are images of order in chaos. The self-similar patterns, the little whirlpools mirrored in larger vortices, are called *fractals*. Fractals are the thumbprint of order in chaos. They indicate structure in an evolving system. They promise predictability. Knowledge about one isolated region in one single whirlpool can be scaled up to capture general rules about the system in its entirety. Fractals are the universe in a grain of sand, and fractal geometry is the tool for discovering chaotic attractors.

"We were astonished when we saw these beautifully organized vortices," Anatol Roshko remembers. "Something spectacular had happened. It was a lifetime experience for us. No one expected it, and we even wondered if something might be wrong with the experiment."

"I couldn't believe what I was seeing," Garry Brown recalls. "I spent the next three months trying to get rid of it. But we eventually had to accept the fact that these patterns are an intrinsic part of turbulent fluid flow. We were looking at the edge between structure and nonstructure in an evolving system. We discovered there is predictability in turbulence, which can be modeled or calculated. Crude assumptions about small-scale structure might allow you to capture the large-scale structure that dominates the mechanics of the system as a whole."

Instead of blowing gas down a wind tunnel, imagine blowing money, all the money in the world, which is tumbling in great rivers of investment capital, Eurodollars, forex cross-rates, index funds, straddles, strangles, floats, and other financial flows that move daily through the world markets. These flows appear at first to be random. They are nothing more than streams of data, masses of numbers as murky as gases wrapped around one another in tessellated braids.

Now send a probe into this data stream, one of the new, nonlinear tools designed for finding order in complexity. This is your high-voltage spark. This is your lens for snapping pictures at the edge of chaos. One good picture would go a long way toward ending the debate about structure in the world financial markets. But one needs multiple probes over multiple time scales in order to build a predictive system out of which one can make money.

"What you really want is a movie that would highlight the transactions of all the traders in the world," says Doyne. "In this movie you'd be able to follow the money flowing back and forth. Money tends to swish in a certain way. From New York to London and out from there to Singapore, the money swirls in vortices that you want to probe for evidence of large-scale structure. The trick is to find the right probes and ways to analyze the data coming off of them. To make this movie, you need a palette of different techniques." The palette includes linear methods, auto-regression, polynomials, nearest neighbor approximations, kernel density estimates, multidimensional histograms, neural nets, and radial basis functions. Once these colors are on your palette, you have to mix them. This task is aided by search algorithms. Here the choices include back propagation, Monte Carlo methods, simulated annealing, conjugate gradients, and trial and error. Common to all these techniques is the fact that they are computer-intensive tools for finding patterns at the edge of chaos, and perhaps even in fully developed chaos, which is where finance takes place.

For most of the people in his audience, Doyne's remarks are Greek. Few Wall Street traders are mathematically advanced, but they appreciate his honesty when Doyne warns them that none of these sophisticated techniques is guaranteed to be free from human error. "There are billions of ways to hallucinate and play tricks on yourself," he cautions. "The landscape abounds with mirages. You build a model. You try it on the data. You don't like how it's doing. You get in your time machine and zip back through the data with the foreknowledge that your previous model didn't work. You don't have to go around this loop very many times before you've burned up your data. You're now hallucinating. You think you have a terrific model, but all you've really learned how to do is predict the past."

Even after Brown and Roshko snapped their famous photo, the turbulence debate raged for another thirty years. Those who believed that turbulence is random argued that Brown and Roshko's apparatus was faulty. They had not really taken a picture of turbulence. They had not really cranked the system as hard as it could go. "Financial forecasting is locked in a similar debate," Doyne remarks. "Efficient markets are supposed to be random. Even if people were confronted with a picture as clear as Brown and Roshko's, the stalwarts of efficient markets would call it an artifact, a fake. Once again, all we can do is wait for the old-timers to die off."

Why is there structure in financial markets? Why does the data cluster in predictable patterns? The short answer is simple. Financial markets are the product of human activity, and humans are irrational, trend-following, herd-driven creatures who react and overreact *en masse*. "I believe in the *in*efficient market theory, based on human foibles and the herd behavior of people acting in groups," Doyne declares. "Traders share common knowledge and biases. We all read the same newspapers. People tend to look at the world in a similar manner. We share common emotions, like fear and greed. We obey universal rules of human psychology and act in herds.

"Laws of supply and demand are moving, psychologically perceived targets," he continues. "Markets depend on the way people think about markets, which in turn influences the way they trade, which influences what the markets do, which influences the way they think about them. There's a kind of regress. We know from control theory that lags between phenomena and the controls placed on them will make things oscillate. A similar lag exists between receiving news and trading on that news. Time lags and delays in market perception create oscillations.

"Another important factor is the positive feedback that exists between perceptions of markets and the markets themselves, which creates self-fulfilling prophesies. If people think the dollar goes up with political uncertainty, then when there is political uncertainty, they will place their bets on the dollar going up, which *will* make the dollar go up. The world is a global poker table, and there is more to playing poker than just the odds; you also have to anticipate the moves of the other players."

Doyne, after delivering his sales pitch at Kidder Peabody, begins hustling around town, visiting other Wall Street firms with fifty million dollars to spare. Next stop is the seventieth floor of the World

Trade Center, where he is invited into the Persian-carpeted office of a New York banker. The banker is a patrician gentleman who manages money the old-fashioned way. He reads the paper; he talks to his friends; and he looks through his plate-glass window onto New York Harbor, which gives him a good sense of which way the wind is blowing.

Doyne's presentation is quite professional by now, especially after coaching from his business-savvy friends, but Doyne is not yet fully transformed from physicist to financier, as he soon discovers. He is trying to mix pleasantries with business. Down below, the harbor is filled with little boats rounding the tip of Manhattan and ferries chugging off to Ellis Island. Doyne decides to skip the theoretical stuff and shuffle straight to the numbers. He hands his host some graphs. On top of the pile is a graph of expected profits to be earned from trading a black-box portfolio of currencies. The graph skittles around in the usual sawtooth fashion, whipsawing through various troughs and peaks, before ending with a triumphant four-hundred-percent profit over five years.

"This is great. These are really good results," says the banker, smiling. "But if you had been working here, I would have fired you, twice." He points to the two biggest drawdowns in the portfolio, where its value dips precipitously before resuming its march to the top.

The predictors had tried to make the ride as smooth as possible, but it still had bumps. While adjusting the dials that control risk and return they had cranked up the leverage, which is the amount of borrowed money in play. Leverage gives higher returns but greater risk, and Doyne knows that conservative investors have a low tolerance for this four-letter word.

The best solution for straightening out Prediction Company's graph is to increase their Sharpe ratio. This statistic measures the risk an investor takes for each increment of gain. Prediction Company's Sharpe ratio is considerably better than that of the S&P Index over the preceding five years, but with the leverage knob turned up so that returns are high enough to suit the high rollers, the Sharpe ratio needs to be higher still.

Sharpe ratios below 1.0 are like sailing in choppy seas. Sharpe ratios near 2.0—the kind preferred by the patrician banker with whom Doyne is talking on the seventieth floor—represent smooth sailing on a sunny day. With a really big Sharpe ratio, up near the double digits, the graph of money versus time looks like a line going straight up to heaven. This is the kind of Sharpe ratio that gives both low risk *and* high returns, the best of all possible worlds.

Back on the ground, Doyne heads uptown to Citicorp. Citicorp is his ace in the hole. Doyne has met John Reed, the chief executive. They spent a day together in New Mexico talking about the Third World debt crisis. Reed was impressed enough with Doyne's ideas to offer him access to the bank's financial data. Now that Prediction Company is in business, surely Reed will want to put his name on their dance card.

As Doyne walks into Citicorp's midtown skyscraper with the chopped top, he notices something strange. The building is quiet as a tomb and many of the offices are empty, as if a neutron bomb has left the desks and paperwork in place, but wiped out the people. The two traders leading Doyne to his meeting stop him in the corridor to whisper that they, and everyone else in their group, have just been fired. To cover billions of dollars in bad real-estate loans, the bank is sacking five thousand people. Doyne addresses the now-defunct group of traders, who sit in stony silence, except for the patter coming from one corporate survivor.

"Your numbers look very good," says the survivor. "You've obviously done a lot of rigorous statistical testing, but no matter how well-tested your results may be, I am going to divide them by two and a half."

"Why is that?" asks Doyne.

"It's just my rule of thumb," he says. "I always divide expected returns by two and a half."

Doyne cuts short the meeting and heads back downtown to Wall Street.

The following day, he is giving his sales pitch at Salomon Brothers, the New York investment bank that had recently been caught cheating its customers in a bond scandal. The chairman has been sacked and Warren Buffet is being brought in to save the company. Doyne is amused by how garish the place is, with oak-paneled rooms, gilt-framed paintings, posh carpets, potted palms. He is eating lunch in the executive dining room, with obsequious waiters hovering around him and all New York spread at his feet.

Salomon Brothers is the company whose traders, calling themselves "big swinging dicks," exemplified the most brazen excesses of Reagan's Roaring '80s. "I had written them off after the bond scandal," Doyne confesses. "But I was surprised to find they weren't the boorish monsters I feared. They were sharp. They asked a lot of bright questions. These must have been the intellectuals in the firm."

The Salomon traders like what they hear, and they want to do business with Prediction Company, but they can't figure out what kind of business. "We told them up front we weren't interested in jobs with Salomon Brothers; so that automatically made us strange to deal with. They had to figure out some kind of profit-sharing arrangement, with a fire wall between their technology and ours. This way neither of us would be giving away company secrets." Four meetings later, Salomon Brothers is still confused about how to structure the deal. They believe Prediction Company is the train to the future. It is roaring down their track, and they want to be on it. But they don't know how to get on board.

"Companies like Salomon Brothers are organized around asset classes," Doyne explains. "They have departments specializing in currency trading, equity trading, and fixed income, which is bond trading. They think in terms of discrete markets, while we were proposing to create something that would cut across markets. We wanted to trade in equities, bonds, currencies—the whole works. At Salomon no one group was willing to fund us if we were going to be useful to other groups." At the end of his fourth meeting, each with a different gaggle of executives, Doyne decides to steer clear of this headless giant.

Doyne and Jim McGill keep shuttling from midtown to Wall Street, riding elevators to meetings at Merrill Lynch, Morgan Stanley, Goldman Sachs. Dozens of busy executives file into conference rooms to hear Doyne's sales pitch. Getting to the top is no problem, but getting over the top is proving tricky. Their visit to Paine Webber is an example. When Doyne and Jim McGill show up at the midtown headquarters of this brokerage firm for the third in a series of meetings, they are ushered into Don Marrin's office. Marrin is chairman of the board, CEO, and president of Paine Webber. He is surrounded by a bevy of eight vice presidents. Untested start-ups without a penny to their name do not usually get this treatment, but Marrin's son has been reading James Gleick's book on chaos theory, which includes a chapter on Doyne and Norman's early research in Santa Cruz. Marrin is jazzed about chaos and the possibility of using it on Wall Street. There are no players in this area, he tells them. You're the only game in town. He doesn't have the math to understand what Prediction Company does, nor do his vice presidents, but Marrin is sure that "chaos investing" is a franchise he wants to buy, and these are the major players. So if he doesn't buy them, someone else will.

McGill mentions that Prediction Company is thinking of signing a letter of intent with Montgomery Asset Management, who want to start a prediction fund. Marrin counters with an offer to supply Prediction Company with operating expenses for a year and money to trade with. Paine Webber has already paid them enough to get Prediction Company to wait thirty days before signing with Montgomery or anybody else. All we want you to do is talk to us, Marrin says.

"The whole thing strikes me as pretty crazy," Doyne remembers. "It's like being paid not to grow potatoes, when you've never grown potatoes before. Marrin wanted chaos and fractals, and we were offering engineering and statistics. So of course he was disappointed when he figured out we were talking about different things."

By the end of Doyne's exploratory probe onto Wall Street, a surprising number of people are willing to open their checkbooks and make a bet on Prediction Company. Among the contenders are Merrill Lynch, Paine Webber, Salomon Brothers, Montgomery Securities, Bank of America, and Kidder Peabody. They represent a thick slice of American capital, with billions of dollars among them. *Not a bad start*, thinks Clem the Cowboy from New Mexico.

Cloning Mr. Jones

Finance is a pure information processing game. A lot of
people in the business are doing things that should be
done by computers.

— DAVID SHAW

We're cowboys in the purest sense.

— PAUL TUDOR JONES

A lapis lazuli sky is draped over Santa Fe's sun-washed buildings. Up
in the hills around the city the leaves on the aspens have turned gold.
Early on this sparkling morning in October the predictors are gathered in the living room of the Science Hut. They sit in the lawn chairs
and mock mission ladderbacks that have begun to crack under the
weight of six-footers rocking back and forth with their feet propped
on the table. The old couch dragged in for Zeus Pelkey is occupied
today by George Cowan, the Los Alamos nuclear chemist who
founded the Santa Fe Institute. A courtly, white-haired gentleman,
Cowan, as one observer describes him, looks like "Mother Teresa in a
golf shirt."

Given the perpetual "dress-down Friday" that pertains on Griffin
Street, the predictors are wearing blue jeans and tennis shoes, or no
shoes at all. They are staring at the company whiteboard, which is
covered with a rash of red equations. Around the equations are those
images still in the running for the Prediction Company logo. The
motto jokingly proposed for the company is *Lux Solis Ex Cucumero,*

Sunlight from Cucumbers, which is borrowed from *Gulliver's Travels*. If cucumbers make juice from sunlight, one should be able to reverse the process and get sunlight from cucumber juice. If the market makes numbers out of information, one should be able to reverse the process and get information out of numbers. The hope is that the world financial markets are more amenable to reverse engineering than cucumbers.

Everyone in the room is listening attentively as Doyne stands at the whiteboard giving a lecture on portfolio theory. "This is one of the most useful things to come out of economics," he says. "The point is to get the highest return for the least risk. Everyone is greedy. They always go for the largest expected return. At the same time, everyone is risk averse. They always go for the least risky investment."

Doyne fills the whiteboard with equations describing how investors teeter between greed and fear. He unfurls more equations describing how portfolios, assembled from different kinds of investments, can damp down risk at no loss to return. Across the board run Lagrange multipliers, sigmas, and the other arcana of mathematized economics.

"I think you reversed the signs in your last equation," Tony chimes in.

"You're right," Doyne says, correcting the error. "The last thing you want to do is maximize your risk."

After shrinking the subject into a forty-minute lecture, Doyne begins speculating on how portfolio theory can be improved. "I'm playing with voting algorithms," he says. "Imagine polling thirty currency traders and acting on their consensus opinion about what's going to happen. I want to set up enough models to do this polling mathematically."

"You might want to look at intrinsic differences between various sectors of the economy," Cowan suggests. "This could be another way to build balanced portfolios." As well as being a respected scientist and administrator, Cowan is also the millionaire founder of the Los Alamos National Bank.

"We have a problem burning economic data," Doyne continues. "There just isn't very much of it in the world. Futures trading on currencies was invented in 1972. Futures contracts on the S&P 500 were invented in 1982. Until recently, only daily data was recorded. They saved opening and closing prices, but not the intradaily ticker-tape transactions. Even macroeconomic data on the economy as a whole is reliable only for the past forty-five years."

Data is *burned* whenever a learning algorithm evaluates a series of numbers and tries to forecast their future evolution. At each pass through the numbers, the algorithm gets smarter at fitting a curve to the data, and with enough passes, the curve will be a perfect fit. But there is also a danger of overfitting the data. In this case, all one has learned to do is "predict" the past—an admirable feat, but useless for a company hoping to predict the future.

"Voting cultures sometimes run into trouble," Cowan advises. "You don't want your algorithms ganging into a block vote."

Doyne continues Cowan's line of thought as he describes how a truly balanced portfolio would cut across all sectors of the world economy, mix every financial instrument, and change dynamically in real time second by second.

"I was up until three in the morning digesting the standard texts on portfolio theory," Doyne confesses. "Economists use a lot of jargon, but once you see what they're up to, it's pretty straightforward."

The meeting ends with everyone drifting off to work. Doyne, to get around the problem of burning data, is creating synthetic stock markets. John is building a confidence assessor for evaluating predictive models. Tom is rigging up an alarm to blow the whistle on overfitting. Tony is phoning Wall Street looking for data. Joe and Stephen are debugging Doyne's trade engine, which is twenty thousand lines of computer code dedicated to converting predictions into buy and sell orders. Norman, off in Illinois, is thinking about the relation between snowflakes and self-organizing adaptive systems, like the stock market. Jenny is regluing the leg on a broken chair. Helen is writing the first five-hundred-dollar-a-week paychecks, which go to the junior researchers. McGill is wandering from room to room reminding everyone that the company's Q4 objective is to "create a functional product."

Doyne walks Cowan to his car, which is parked in the back alley. They are standing by the garbage cans when Cowan mentions he might like to invest in Prediction Company. Doyne politely declines. Rather than mortgaging the business to venture capitalists, he has a better idea. He and McGill are soon winging their way back to Wall Street, where they have appointments with two of the city's smartest entrepreneurs. One is an expert in artificial intelligence who jockeys a hundred-million-dollar investment in computerized trading programs. The other is a "cowboy" who beats the market by donning a pair of lucky tennis shoes.

David Shaw's ambition is to put the New York Stock Exchange out of business. "Many of these people are doing something that should be done by computers," he claims. "It's just not very complicated. All you have to do is match a buyer and a seller. For serving this middleman function, you pocket some money. The middlemen often have a monopoly franchise on a given stock, which they inherited from their families. These franchises, quite literally, are passed down from generation to generation. It's basically a license to collect exorbitant amounts of money for a very simple process."

"I find a lot of finance highly amusing," says Shaw. "It's hocus-pocus practiced by people in fancy suits. Quite often they're selling you financial products that are terrible deals or providing very mechanical services at inflated prices." Once computers have replaced the tassel-loafered leeches on Wall Street, "these people can try to educate kids in the inner cities or fix the manufacturing sector or perform other value-added services," he adds.

David Elliot Shaw is eponymous head of the New York investment bank D. E. Shaw & Co., which manages a billion-dollar hedge fund. D.E. Shaw trades enough shares on a busy day to account for five percent of the total volume of the New York Stock Exchange. It ranks among the top twenty securities companies in the United States. It is a major player in the Japanese markets and employs four hundred people in offices from London to Hyderabad. In 1996 Shaw started giving away free e-mail accounts through a company called Juno, and in 1997 he launched FarSight, an Internet-based service for one-stop banking and stock trading.

A gangly, flat-footed man in his forties, Shaw is a former Los Angeles surf musician who parlayed his knowledge of computers into a financial empire that *Fortune* magazine describes as "the ultimate quant shop." *Quant* is short for *quantitative analyst*. This computerized mix of mathematics and statistics is also known as *rocket science*. But Shaw knows the difference between real rocket science and Wall Street's version; so his ears perk up when he gets a phone call from someone fresh off the hill at Los Alamos. "We'd like to see you as soon as possible," Shaw's assistant tells Doyne.

Born in 1951, Shaw grew up in the Los Angeles neighborhoods around UCLA, where his stepfather, Irving Pfeffer, was a professor of finance. Shaw's natural father (his parents divorced when he was twelve) was a theoretical physicist who studied plasma and fluid flows. Shaw later combined these two influences to become a proponent of phynance—the marriage of physics and finance. Phynance studies money as a kind of fluid flow. It searches for little eddies of predictability in the great waves of speculative investment that wash daily around the globe.

Shaw first began thinking about these problems in the late 1960s, when he accidentally found himself taking a college course in cognitive psychology taught by David Rumelhart at the University of California at San Diego. Rumelhart is one of the pioneers in the development of synthetic neural nets. Neural nets are computer systems composed of simple, interconnected processing elements. These simple elements, resembling neurons in the brain, are capable of performing complex computations, recognizing patterns, and displaying other forms of artificial intelligence. Shaw later went to graduate school in computer science at Stanford, where he set out to do for artificial intelligence what the supercomputer designers in Silicon Valley were doing for number crunching. He asked himself, "What if we throw out the ground rules and design a completely new type of computer, whose architecture is suggested by the brain? Nobody has ever built this kind of computer before."

Shaw was one of the first people to design and try to build a massively parallel computer. Rather than the normal, digital computer, which moves sequentially through data, one step at a time, Shaw's prototype machine, which looked like a tree leafed out with thousands of computer chips, took a multipronged approach to problem solving. It employed thousands of synaptic processors that fired all at once. The full-blown version would be awesomely fast. It would also be awesomely expensive, with a price tag of a hundred million dollars.

Shaw had moved to New York in 1980 as an assistant professor of computer science at Columbia University. He was shopping his idea around Wall Street, having no luck finding investors, when he got a phone call from a headhunter for Morgan Stanley. They had heard that Shaw was a computer whiz and that he was good at running big projects. These skills were in short supply on Wall Street in the mid-1980s, where the major banks and investment firms were desperate to junk their mainframe computers and boost their computational intelligence. The long-haired, bearded, Jewish antiwar activist from southern California had qualms about working for a white-shoe Wall Street firm, but Morgan Stanley offered to put an extra 0 on his salary, and the project looked interesting. "What appealed to me about the job was the challenge of trying to beat the market," says Shaw. "I was raised to believe it was impossible, and here they were telling me they knew how to do it."

In 1986, Shaw was hired into the Analytical Proprietary Trading group at Morgan Stanley. This was a top-secret, "black-box" operation that worked behind a guarded door on the nineteenth floor of the Exxon building in midtown Manhattan. They were trying to exploit anomalies in stock market prices. Their job—aided by Shaw's computational expertise—was to isolate recurring patterns in financial data and use them to beat the market. The group was run by Nunzio Tartaglia, a former Jesuit seminarian with a Ph.D. in astrophysics, and Bill Cook, who was head of information technology at Morgan Stanley. It was Cook who later founded Tech Partners and tried to lure Norman Packard to Wall Street in 1991.

The secret being exploited at Morgan Stanley was something called pairs trading. Pairs trading traces its lineage back to Jesse Livermore, the famous speculator whose life story is recounted in *Reminiscences of a Stock Operator*, first published in 1923. Pairs trading is based on the idea that prices of related stocks should be correlated. Ford and General Motors will tend to fluctuate in price around the same news events. But what if an unusual gap develops, where Ford lags in price and GM pushes ahead of their normal relationship?

In this case, a stock speculator might go short, or sell, GM and go long, or buy, Ford. The market can crash. The market can soar. But if the gap between the two stocks remains unchanged, our speculator will neither win nor lose money. Such strategies are called *market neutral*. The bet is being placed not on which direction the stock market will move, but on company-specific or sector-specific correlations. If the gap between GM and Ford narrows, with GM falling and Ford rising in price, as predicted, our speculator will make money. How much money? One can only guess, but if Morgan Stanley was hiring astrophysicists and tacking additional 0s onto the salaries of college professors, the strategy must have been rewarding.

A key trick to statistical arbitrage, or *stat arb*, as Wall Street refers to pairs trading and other forms of algorithmic hedging, is to get one's strategy truly market neutral, not only to fluctuations in the stock market, but also to fluctuations in interest rates, foreign exchange movements, and global economic risks that can come at you faster than a shark going for red meat. Wall Street is filled with former employees of "market-neutral" hedge funds that went belly up because they weren't neutral enough.

Shaw worked for a year and a half at Morgan Stanley before striking out on his own. He wanted to expand pairs trading to include multiple classes of securities—bonds versus stocks versus options—and multiple markets hedged against one another in a computer network designed to be *global* market neutral. Shaw was going to *stat arb* the world; he was going to hedge the universe. Today, his computer models are so powerful that Shaw, along with locating market inefficiencies, also detects rigged markets and other forms of cheating. "It's an everyday, predictable cost of doing business." He shrugs. "We know that trading with other brokers is invariably worse than trading with individual customers."

Stat arbs do not forecast major market moves; they hedge against them. Prediction Company is taking a different approach. It hopes to accomplish what most professional investors say is impossible: foretell the future. This is why Shaw is intrigued by Doyne's presentation. It encompasses vast new realms of information that he is hungry to process.

Shaw is a personable man, an adviser to the President on computer technology and a supporter of good liberal causes, but he is also a control freak capable of flying into outsized rages. He rarely lets visitors inside his company, and he even keeps his own traders in the dark about how the firm's computer models work. "We follow the same principles as the CIA or NSA," he says of security measures at D. E. Shaw. "Information is partitioned off and distributed on a need-to-know basis." When asked by a reporter if he used neural networks to design his trading system, Shaw jokingly replied, "I could tell you, but then I'd have to kill you afterward."

When Doyne flies into town to meet him, Shaw is in the process of moving his company into new offices on the top two floors of a sky-scraper near Times Square. The space is decorated with high-tech "mushroom" chairs, shiny black floors, and cutout walls bathed in reflected light. D. E. Shaw's logo is an electronic switch, and the offices are meant to evoke the feeling of sitting inside a computer chip. Shaw's trading room is a computer-intensive, black hexagonal chamber that makes the *Challenger* space capsule look Early American. When he shows up for work, which is usually not before noon, Shaw sits at a brushed aluminum, wing-shaped table ergonomically designed so that his neck swivels no more than fifteen degrees between computer monitor and desk surface.

Doyne finds the mushroom chairs incredibly uncomfortable. The paint fumes give him a headache. He is called back for two, three, four meetings, and the negotiations are not going well. Shaw seems to be offering to finance Prediction Company for two years, but then he wants to own their technology. Doyne, not liking the proposal, decides to call it quits.

Paul Tudor Jones II is rocking through a go-go day, making a killing in T-bonds while simultaneously shooting ducks in the commodities markets. He is unwinding million-dollar positions faster than the cuts on an MTV video clip. Wearing the lucky, high-top sneakers that once belonged to actor Bruce Willis, Jones is a dervish dancing around his office. His voice is hoarse from shouting phone orders down to the New York and Chicago trading floors. "Rock and roll!" he yells to the data screens stacked on his desk.

"Buy one thousand silver at forty-one," he calls to his floor broker in Chicago. "Don't show your position."

"Buy one hundred gold. No!" Jones corrects himself. "Buy one *thousand* gold."

Jones is surrounded by video monitors and speakerphones connected to the four brokers he keeps in every major market. He sings out the numbers. He flails his arms like a conductor calling up the woodwinds. When something on the news feed catches his eye, he starts pacing in front of the screens. Now he looks like a basketball player, twirling on his heels and moving fast for the layup.

"Buy three hundred at thirty!" he yells to a clerk in Chicago who is standing near the S&P pit. "Go, go, go! Are we in? Did we get it? Talk to me, Billie. Talk to me!"

"Yea, we're in," Billie shouts back.

"Do a thousand now, but don't show size," Jones admonishes. Then he suddenly changes direction.

"Kill the second bid!" he hollers. "Bank stocks are cratering. Back out of it, Billie. Back out!"

"You're out," Billie reports, before the line goes dead.

Jones, who resembles a dimple-chinned Clark Kent with a receding hairline, is a former trader from Memphis, Tennessee, who made his first fortune on the floor of the New York Cotton Exchange. Afraid he was losing his voice, he left cotton trading to make a second, much grander fortune running Tudor Investment Corporation. He has offices in London and Tokyo and occupies a choice piece of lower Manhattan real estate at One Liberty Plaza, which has stunning views over the city, and every amenity, including a gym full of StairMasters. Jones also rents a dingy office across the street. This is where he got his start as a cotton trader, and he doesn't want to forget where he came from.

Jones entered the pantheon of Wall Street traders in the 1980s, when he more than doubled his money every year for five years in a row. This included a pleasant 529 percent profit in 1984. Jones really hit it big after the stock market crash of October 19, 1987. After Black Monday his fund went up sixty-two percent for the month of October alone. Jones accomplished this feat by selling stock index futures—a bet the market will go down—and buying Treasury bonds, which is where terrified investors flock for shelter when the market crashes. By the end of 1987 Jones had cleared a personal profit of between eighty and one hundred million dollars. He topped that year's list of Wall Street moneymakers, beating out George Soros and Michael Milken, who held the number two and three positions. Jones at the time was thirty-three years old.

He commutes to his office by helicopter. He owns a fleet of planes for flying down to his Chesapeake Bay estate in Virginia. His twenty-room mansion is surrounded by an award-winning garden with twenty thousand tulips, an aviary, and a petting zoo. Across the bay in Maryland, Jones maintains a three-thousand-acre wildlife preserve, where his duck blinds are equipped with computer screens and telephones. Even Jones's charities are outsized. He has promised to pay the college tuition of a hundred students in a run-down neighborhood in Brooklyn, and his Robin Hood Foundation take millions of dollars from the Wall Street rich to give to the New York poor.

"I attribute my success to the Elliott Wave approach," says Jones, who also believes in lunar cycles, Fibonacci numbers, and other voodoo pulled from the magician's trunk known as technical analysis.

Wall Street investors divide into two camps. There are fundamentalists and technicians. Fundamentalists pore over economic data and corporate reports, trying to assess "value." Value investing is opposed to fads. It is buttoned-down and Protestant in its rigor. It is the steady, sober business of looking for long-term rewards. Technicians, on the other hand, are more inclined to believe in divine intervention. The markets, which work in mysterious ways, can suddenly reverse fortunes and endow even the most downtrodden stocks with heavenly returns. Technical traders decorate their offices with charts and diagrams and are catholic about embracing any approach to trading that seems to work.

Fundamental investing tries to predict the future prospects of a company, primarily earnings. This future value should be reflected in the price of the company's stock. Fundamentalists make money by buying "undervalued" stocks and holding them until the marketplace realigns price and value; in other words, until the stock price rises. The world's best-known proponent of value investing is Warren Buffett, the Omaha stock picker who ranks behind Microsoft's Bill Gates as America's second richest man. Buffett rents a sports arena in Nebraska to hold his yearly shareholder's meeting, which he calls a "capitalist's version of Woodstock." As for *his* trading secrets, Buffett says he learned everything he knows about stock picking from a book first published in 1934 by Benjamin Graham and David Dodd called *Security Analysis*. Buffett got an A+ from Graham when he took his course at the Columbia University Business School, and he briefly worked for Graham on Wall Street before going into business for himself.

It is hard to criticize someone as rich as Buffett, because he is obviously doing something right, but economists who believe in efficient markets dismiss him as nothing more than a lucky monkey. If the markets are efficient, continually fine-tuning themselves to available news, then there is no such thing as an "undervalued" stock. For every Buffett who makes a killing in the market, there are thousands of would-be Buffetts getting wiped out by confusing fundamental analysis with fundamental ignorance.

For his part, Buffett scorns the "egotistical orangutans" who believe in efficient markets, which he calls the "coin-flipping" school of market analysis. In a speech entitled "The Superinvestors of Graham-and-Doddsville," which Buffett delivered at a seminar celebrating the fiftieth anniversary of the publication of *Security Analysis*, he explained that the stock market is not a "national coin-flipping contest." It is actually an arena where plenty of opportunities exist for investors to exploit "gaps between price and value." Buffett went on to describe himself as inhabiting "the very small intellectual village of Graham-and-Doddsville," where there is "a concentration of winners that simply cannot be explained by chance."

Buffett's "very small intellectual village" is now a national fad, but one should note that the founder of Buffett's village eventually moved out. When he was interviewed shortly before his death in

1976, Benjamin Graham, whose original family name was Gross-baum, said he had undergone a change of heart. He no longer believed in the kind of security analysis geared to finding stock-picker's bargains. "That was a rewarding activity forty years ago, when Graham and Dodd was first published, but the situation has changed," he said. "Today I'm on the side of the 'efficient market' school of thought."

Technical analysis cares nothing for "value" or other economic funda-mentals. It claims that everything one needs to know about a com-pany's stock performance can be read from the "tape"; that is, from the history of past prices and the volume of trading at these prices. Both fundamentalists and technicians can use pattern recognition or build models to forecast market trends, and Prediction Company relies on both kinds of data. But technical analysis, as it is generally practiced today, is a mystical enterprise that has more to do with alchemical incantations than science.

Elliott waves are one example. This arcane theory of market cycles is named after Ralph Nelson Elliott, an anemic accountant who fell into a coma during the stock market crash of 1929 and woke up later during the Depression, curious to know where his money had gone. Elliott's answer, published in 1946 under the self-effacing title *Nature's Law: The Secret of the Universe*, remained as obscure as Elliott's other major publication, *Tea Room and Cafeteria Manage-ment*, until Elliott's ideas were resurrected in the 1980s by the stock market guru Robert Prechter, who publishes something called the *Elliott Wave* newsletter out of Gainesville, Georgia.

Elliott believed there are predictable waves of investor psychology that drive the market through its upswings and downturns, and he developed an elaborate taxonomy for predicting these market shifts. The mother of all Elliott waves is the two-hundred-year Grand Super-cycle. Inscribed inside this big wave are eight smaller waves progress-ing down through the Supercycle, Cycle, Primary, Intermediate, Minor, Minute, Minuette, and Subminuette. At this very moment, hundreds of chartists, employed at Wall Street's most prestigious firms, are trying to figure out which wave the world economy is currently riding. Before he stumbled in 1987 and went from being the best- to the worst-performing analyst on Wall Street, Prechter could make Elliot-tic pronouncements that moved the Dow Jones Industrial Average a dozen points and produced billions of dollars in gains or losses.

The grab bag of technical analysis includes Fibonacci numbers, a numerical sequence originally used by a thirteenth-century Italian mathematician to explain why rabbits breed like ... rabbits, and Kondratieff waves, named after a Russian economist who was sent to Siberia by Stalin for believing that capitalism, rather than being doomed to failure, actually rises and falls through long-term cycles. These methods are probably no more useful than astrology, but this, too, is a major force on Wall Street. A former employee at Merrill Lynch publishes an astrological newsletter devoted to market timing. The newsletter, which is also published in a German edition for bankers in Zurich and Bonn, is mailed to a thousand subscribers. This is part of the growing interest in a field known as "mystical investing." The financial astrologers who attended a recent conference on Wall Street controlled more than four billion dollars in funds.

"It's like the Middle Ages again," says Doyne. "On one side are the alchemists holed up in their caves with their Elliott waves and magic spells, searching for incantations to foretell the future. Many are char-latans or misguided numerologists. But that doesn't mean all of them are. On the other side are efficient market theorists, who remind me of the Pope in the fifteenth century pronouncing that the world is flat and Copernicus is wrong—a position the Church maintained until 1828."

It is obvious to Paul Jones and other technical analysts that financial markets move in cycles and that the trick to beating them is to predict these cycles with enough forewarning to outrun the crowd. "I believe the very best money is to be made at market turns. I have caught a lot of bottoms and tops," says Jones, who is not averse to making what he calls "Nostradamus trades." Named after the sixteenth-century French astrologer who predicted worldwide famine in the Year of the Comet, these trades are keyed off full moons, El Niño weather pat-terns, comets, and other planetary events.

Jones says he "scripts" markets, trying to figure out the cyclical and sentimental factors that make them fluctuate. He keeps thirty markets in his head at one time, in the form of visual diagrams. These diagrams speak to him with the telltale signs that chartists use to predict market turns. "Suddenly a lightbulb clicks on and you start visualizing in advance the script the markets are following. These are the occasions that you have to max out with mega-positions and play just as hard as you can stand it."

Whether or not his method works, if enough people are "scripting" markets with Nostradamus trades, these patterns will become self-reinforcing. This phenomenon was noted by the great British economist John Maynard Keynes, who likened the stock market to a British beauty pageant. The winner is not the most beautiful woman, but the woman whom most people think everyone else will choose as the most beautiful woman. "In estimating the prospects of investment, we must have regard to the nerves and hysteria and even the digestions and reactions to the weather of those upon whose spontaneous activity it largely depends," wrote Keynes. He used this knowledge to make his own fortune in the markets. But unlike Jones, who married a New York model, Keynes married a Russian ballerina.

While Paul Jones and Doyne are eating lunch in his office, Jones keeps an eye cocked on the video screens hung throughout his building. Once he reaches for the phone and barks an order to a trader in Chicago to "dump ten thousand SPUs." The conversation resumes where he left off.

Jones's trading follows the sun. He plays the commodities and futures markets during the day. In the early evening he hits the late bond session in Chicago. He checks the currency and gold markets before going to bed, except when the markets are hot, and then he fields calls all night long, as the book moves from Tokyo to Singapore and on to London at three in the morning. Jones is a success. He is also a prisoner of his success. Even on vacation, he has to travel with a trunk full of electronic gear and stay constantly tuned to the markets.

In 1989 he took the unusual move of returning two hundred million dollars to his investors, saying his fund was getting too unwieldy.

Rather than playing the market, he was *becoming* the market. At the same time he hit on the idea of cloning himself. He would do a brain dump and reverse engineer his intuition into a computer program capable of replicating his market moves. He would educate the computer with trading rules and endow the machine with a learning algorithm that allowed it to get smarter and smarter at being Jones, until one day it became Superjones, smarter than the master, but never tired or longing for a vacation.

Jones hired a team of financial engineers. He put together one of the finest research organizations in New York and assembled the best database in the country. He was in the midst of spending nine million dollars building an artificially intelligent, machine-based Jones called *Madonna*, after his favorite singer, when Doyne dropped by for lunch.

At an earlier meeting with Jones's research staff, Doyne was surprised to find that the ten people in the room had apparently read his and Norman's published papers on chaos and forecasting. "We've implemented all your stuff," they told him.

"I had the weird sensation of walking into some kind of club or fraternity into which I had already been secretly inducted," Doyne says. "I had never met these people, and they were telling me my ideas were already in their machine."

Soon Jones's director of research and five executives are flying the company jet to Santa Fe. Their visit is preceded by a flurry of activity on Griffin Street, as Prediction Company glues the legs back on their chairs and replaces most of their folding tables with regular office furniture.

Jones's team is impressed by what they find. "Compared to the Volkswagen we've built in New York, Prediction Company is a Ferrari. But the problem is, you guys are running your machine on regular, and we're running on jet fuel."

This can be remedied by filling Prediction Company's tank with Tudor's capital. They offer to finance a Prediction Company investment fund and split the proceeds fifty-fifty. It looks as if Prediction Company has a deal, and everyone is excited about the prospect of getting this Ferrari out the door and watching her *go*!

The Perfect Girlfriend

The money is always there, but the pockets change; it is
not in the same pockets after a change, and that is all there
is to say about money.
— GERTRUDE STEIN

It is not easy to get rich in Las Vegas, at Churchill Downs,
or at the local Merril Lynch office.
— PAUL SAMUELSON

Doyne receives a phone call one morning from a man with a broad
Texas accent, who introduces himself as "Tom Dittmer's point dog."

"Tom was thinking of flying from Chicago to Aspen today," drawls
the Texan, "and he'd like to divert his jet down to Santa Fe to visit
you boys."

"He can drop by any time," says Doyne. Then as an afterthought,
he asks, "Who is Tom Dittmer?"

Dittmer is a legendary Chicago commodities trader who used to
celebrate big wins by giving away Rolex watches and flying his friends
to Las Vegas for the weekend on a chartered 747. After a couple of
suspensions for securities infractions, Dittmer left the pits to start
running Ray E. Friedman & Co., known as Refco, which he built into
America's largest futures and options broker. Among Dittmer's clients
is Hillary Rodham Clinton, soon to become First Lady. Mrs. Clinton,
trading cattle futures through Refco, parlayed a one-thousand-dollar
investment into a one-hundred-thousand-dollar return, thereby securing a tidy ten-thousand-percent profit. Dittmer is executing one out

of every five contracts traded on the Chicago exchanges and is recently installed on the Forbes Four Hundred list of richest Americans when his Lear jet touches down in Santa Fe.

Dittmer is one of a stream of high-net-worth individuals—as the rich prefer to call themselves these days—who begin diverting their jets into Santa Fe once an article about Prediction Company hits the front page of *The New York Times*. The article, "From Swords to Plowshares," which runs with a photo of Doyne working at his computer, is one in a series of pieces discussing what the military-industrial complex will do with itself once the Cold War is over. *The New York Times* suggests that Los Alamos National Laboratory, the nation's preeminent weapons research facility, home of the atomic and hydrogen bombs, with thirteen thousand employees and a yearly budget of eight hundred million dollars, should emulate the hang-loose hackers down at Prediction Company. They exemplify what scientists can do when making love not war. They are the avatars of the future, leading the way into the peacetime economy. Soon after the article appears, in February 1992, Prediction Company fields more than a hundred phone calls from potential investors.

The president of a big oil company drops by for a visit. He tells Doyne he is selling his oil tankers and getting into money. As the oil man is walking out the door, up pulls a car full of sober-suited brokers from Merrill Lynch, the world's largest investment company. The brokers have barely come and gone when into town flies Brad Rotter, a former blackjack card counter who now runs a commodities trading company in San Francisco. Over margaritas and tapas at El Farol on Canyon Road, Rotter tells Doyne that Prediction Company is inventing the next Black-Scholes model, and if they play it right, the company could earn five hundred million dollars in the next five years.

Discovered in 1973 by the mathematician Fischer Black and the economist Myron Scholes, who were then teaching at the University of Chicago, the Black-Scholes option pricing model is the preeminent example of an equation that was used, and is still used, to make money in the financial markets. The model gave Wall Street a complicated but workable solution to what until then had been an intractable puzzle: how to value an option. Black and Scholes showed that while stock and option prices vary randomly, they are related to each other via an equation. This equation looks familiar to physicists because it is the same one they use to describe the diffusion of heat. The Black-Scholes model was published the same year the options exchange opened in Chicago, and it played a big part in allowing the derivatives markets to flourish. By giving a scientific imprimatur to these risky investments, the Black-Scholes model was one reason why pension funds, which previously had been banned from the options market, were allowed to jump into the game.

Another former gambler showing interest in the predictors is Blair Hull. Hull made half a million dollars in the 1970s as a member of a team of professional card counters. Ken Uston, "the big player" on the team—he later published a book called *The Big Player*—had had an unpleasant encounter with the Mafia and was having reconstructive surgery done on his face when Hull thought he should look for another line of work. He took his blackjack winnings and started playing the Chicago options market, which turned out to be far more lucrative than the games in Las Vegas. By 1991 Hull's stake had grown to ninety million dollars, and he was running a highly disciplined, computer-intensive operation with a hundred employees.

Some of the potential suitors lining up outside the Science Hut are rich and smart, and some are rich and not so smart, and others are "snake oil salesmen," as Doyne calls them. But he appreciates their dropping by, and each visit adds to his anthropology of Wall Street. "When we started out, I didn't know what a futures contract was. I didn't know what an option was. I didn't know where these things were traded. I was up front about my ignorance, because I figured there was no reason I ought to know these things. We were accumulating information. We were on a fishing expedition."

One thing the visitors keep emphasizing is the sheer size of the financial markets. Speculators swap more than a trillion dollars a day in foreign exchange. This is fifty times larger than the entire output of the American economy. "Once you hear these numbers," Doyne says, "you realize that even a small advantage can allow you to make a huge amount of money."

This is why the people visiting Prediction Company don't care if the predictors know what a futures contract is. Do they know what the *future* is? That's the trillion-dollar question. Do they have an edge, some technology that gives them a chance, even a hair-thin chance, of beating the game? Finance is not like physics, where bombs either explode or they don't. In the financial markets, which have generally been a random, unpredictable game, even the slightest advantage would be a breakthrough. "At this point we were getting jazzed on the large numbers people were tossing around," says Doyne. "Even if we didn't believe it for a second, there's an undeniable adrenaline jab that comes from someone telling you you're about to make five hundred million dollars."

The predictors allow themselves to start fantasizing about being rich. They are surviving on graduate student wages or no wages at all. Doyne's bank account has been ransacked, and they have yet to make a penny, but they expect to be high-net-worth individuals themselves quite soon. In their customary, hyperrational way they begin computing how much money one needs to be really rich. The figure is elusive, depending on one's needs and tolerance for risk, but they think it falls somewhere between four and a hundred million dollars.

The predictors also investigate the metaphysical and moral issues involved in being rich. "Since most of the rich people in the world are assholes, what is the secret to being rich without being an asshole?" Joe wonders. "Frankly, I have problems morally with what we're doing. The markets are a zero-sum game. For us to be winners, somebody else has to be a loser. We'll be taking money from orphans and widows; they move too slowly to take advantage of what we know. We won't be helping companies grow. We'll just be siphoning off money for ourselves."

Norman's wife, Grazia, whose background combines socialist militancy with a Catholic distaste for *rentiers*, usurers, coupon clippers, and other free-market parasites, is particularly dour about the company's prospects. "Making money was never a priority for my father," she says. "Creating jobs, not playing games with other people's money, was what drove him as an entrepreneur."

Norman tries to counter Grazia's argument with a spirited defense of how Prediction Company will damp out the oscillations in the markets and actually make them *more* stable. Stable markets are good for companies who want to borrow money to build factories and put people to work. "I'm not sure who we'll be taking money from," he admits. "But instead of widows and orphans, I think it's more likely to be short-term speculators."

Doyne takes a different tack. At a time when money is drying up for pure research and jobs for physicists are disappearing, Prediction Company is giving a dozen people work who might not otherwise have it. "I'm neutral on the question," says Letty, with lawyerly sagacity. She is the only one among them who actually has money, and she knows how to sleep at night without worrying about it. She also knows that Doyne has not received a paycheck for eight months.

While Prediction Company is being courted by high-net-worth individuals, morale back on Griffin Street hits rock bottom. One of the problems is love, or lack of love, or sex, or whatever it is that makes men confined in narrow spaces without the ameliorating influence of women get tetchy. Helen Lyons, the company accountant, is still provisioning them with pastries from Café Fritz. The predictors cheer up when munching her chocolate and raspberry tortes, but then quickly descend back into love-sick moping. The unavoidable fact is that five unmarried males have either (a) no girlfriend or (b) no girlfriend for more than a thousand miles in any direction, and most of these long-distance relationships are on the rocks.

After teaching his last course in Illinois, Norman moves to Santa Fe in January with a strep throat and a temperature of a hundred and three. He has no health insurance, and Grazia is pregnant with their second child, due in June. Nervous about leaving a tenured job for the vagaries of life in the Science Hut, Norman entertains the idea of having Prediction Company give the university a fifty-thousand-dollar grant to hold his job for him. The idea is killed when no one but McGill supports it. "Either you're in or you're out," the junior partners tell him.

Another problem affecting morale is the fact that Prediction Company has yet to find a partner. There has been a lot of dating and no shortage of proposals, but none from Mr. Right. The company is rejecting offers from Lone Rangers and other rich people whose money comes with too many strings attached. Citicorp is on the ropes. Salomon Brothers is a headless giant. Bank of America and Paine Webber lack the trading expertise the company needs. Shaw is erratic. Paul Tudor Jones has changed his mind. Instead of forming an alliance with Prediction Company, he wants to buy them. He is valuing the company at two million dollars and offering to put everyone on the payroll. "Sorry, we're not for sale," Doyne and Norman inform him.

Doyne has made five trips to New York. Norman has shaken the money tree in San Francisco. McGill has spent the fall and early winter living on airplanes. "I'm totally sick of negotiating. It's an intellectual striptease," Doyne complains. "We do a little dance where we reveal enough about Prediction Company's strategy to make it tantalizing, but not enough to give away company secrets."

McGill alone among the predictors is cool and unflappable. He relishes the psychological jujitsu. His only word of solace to the troops back on Griffin Street is a reminder that they have to file both their quarterly and yearly objectives. "Yearly objectives?" mutters Tony. "I don't think we can pay the rent next week."

To put an end to their misery, Doyne and Norman decide to double back to Montgomery Asset Management in San Francisco, which is still gung ho about launching a prediction fund. Prediction Company is not gung ho about schmoozing fund managers in Nebraska, but for want of a better deal, they are about to sign the Montgomery contract when Doyne gets a phone call from Esther Dyson. She is inviting him to be a luncheon speaker at her yearly computer conference in Tucson. Computer nerds in Tucson hold no more interest for Doyne than fund managers in Nebraska, and he is about to decline her invitation when McGill intercedes. "Bill Gates will be at the conference. All the major players in Silicon Valley will be at the conference. You need money. They have it. You're going."

"Okay, I'm going," Doyne agrees.

Esther Dyson is a diminutive speedball with the knowing, brown-eyed gaze of her father, Freeman Dyson, who is the most famous physicist currently installed at the Institute for Advanced Studies.

Dyson *fille* edits a computer-industry newsletter called *Release 1.0*, and she hosts the annual PC Forum, where big shots in the business rub shoulders with one another. She is to the computer industry what Moody's is to bonds. When she says you're good, you're golden. So Doyne writes a speech on the new sciences of prediction—something light enough to amuse five hundred people masticating their way through a low-calorie chicken lunch—and reports to a posh resort outside Tucson.

The speech goes well. There are lots of questions. Doyne has fielded the last of them when up to the podium walks a weedy fellow with spectacles. He is the kind of IT nerd who wears suits instead of pocket protectors and who sports a title like "chief technology officer." A person like this might direct a staff of two thousand Dilberts beavering away in cubicles. The chap introduces himself as Craig Heimark and hands Doyne his card. He is the partner in charge of technology at O'Connor & Associates, which Heimark describes as a major player in the Chicago financial markets. "You ought to be our partners," he says. "I understand you're about to sign another deal. But before you do, you should come talk to us."

By coincidence, an old roulette buddy of Doyne's has sent him an article about O'Connor, with a note saying, "This is who you should be partners with." Doyne has just returned to Santa Fe when he gets a phone call from Chicago. "Hi, this is David Weinberger," booms a New York voice with money behind it. Weinberger is managing director at O'Connor, or something like that. He speaks too fast for Doyne to catch it all. "If you're still looking for a partner," Weinberger offers, "I'd like to fly down and visit you in Santa Fe. Today is Friday. What if I stop by tomorrow?"

Weinberger, a trim, athletic man in his mid-forties, with salt and pepper hair and the nervy assurance of someone who has mastered every game he has ever played, walks into the Science Hut with a couple of "technical people" traveling in his wake. Weinberger has a doctorate in mathematics. He is a former research scientist and college professor who became a big trader at Goldman Sachs, before leaving Wall Street for Chicago. Ideas pop out of him like shards from a fragment bomb. He speaks in loops and cul-de-sacs, his sentences going faster and faster as he tries to get the world packed into one long equation. *The knobs on this guy are turned all the way up,* thinks Doyne.

The technical people are equally quick, and one of them, Albert Zisook, is an old friend of Norman and Doyne's. They worked together on chaos theory when Zisook was employed at Xerox's Palo Alto Research Center. Norman and Zisook found themselves together again in the early 1980s when they both had NATO postdoctoral fellowships in Paris. If O'Connor is hiring scientists like Zisook, then it must be a serious place.

From Weinberger's rapid-fire rundown, Doyne gleans that O'Connor is a *very* serious place. It has six hundred employees shuffling billions of dollars through the futures, options, and currency pits. It is Chicago's biggest dealer in derivatives and "risk-neutral" positions, which are hedged portfolios designed to make money whether the markets go up or down. O'Connor made millions of dollars by being the first company to get the Black-Scholes pricing model to work in the options market. It is the most mathematically astute and computer-intensive broker-dealer in America, and it would make a good partner for Prediction Company. "You should come visit us," says Weinberger. "I think you'll like what you see."

On March 12, 1992, Norman dons the Prediction Company suit and flies to Chicago. He is doing the company "striptease," while Doyne is in Perth, Australia, for a month, delivering some long-promised lectures on chaos theory. After penetrating the electronic lockboxes that guard O'Connor's doors, Norman is shown to a balcony overlooking their trading floor. Down below are hundreds of people clustered into octagonal pods that stretch across a room as large as a football field. The scene is lit by enormous, arched windows facing up LaSalle Street. The room is buzzing with the feral juice that comes off traders in mid-play. It is also buzzing with the electronic juice that comes off computer monitors stacked three deep. Next to boxy Sun workstations, Norman notices the sleek black cubes of the Next computers that are his own favorite machine. The computers are being jockeyed by people wearing blue jeans and T-shirts. This could be a university computer center or a laboratory for designing airplane wings, except for the currency traders who are shouting into their speed-dialers, the light board flashing market news across the wall, and a huge spread of free food laid out for traders too busy to leave the floor.

Norman is ushered into a conference room with windows looking onto O'Connor's electronic ticker tape. The room is soon filled with a big slice of the company's partners; they keep one eye on Norman and the other eye cocked on the tape. Many of them are MIT graduates, including David Solo, a speedy fellow in his twenties who is engineering a computerized system for trading government bonds. Next to Solo is another MIT grad, Clay Struve, a mustachioed, ruddy-faced man who is drinking two diet Cokes simultaneously.

A chunky, fair-haired fellow in his early thirties, Struve is a former rugby prop and mathematical whiz kid who finished college in a couple of years and started working full-time at O'Connor in 1979. Back then the company consisted of a handful of people trying to figure out how to play the options market with the Black-Scholes model. Struve was Fisher Black's student at MIT, where Black had moved after Chicago. He discounts what he did for Black. "Nothing more than odd jobs," he growls. But the fact of the matter is that the student was more adept than the master in applying what he learned. It was O'Connor who made millions of dollars using the Black-Scholes model, not Black or Scholes or lots of other people who didn't know how to turn the numbers into a business.

Struve, who sits in meetings like a speechless, all-knowing sphinx, is the O'Connor partner in charge of risk. The company plays the derivatives markets with leveraged portfolios whose face value is many billions of dollars. Get the numbers wrong, and you go out of business. Struve keeps O'Connor's numbers between his ears. He looks like a hard-drinking frat boy in a rugby shirt, but he has a gift for doing large calculations in his head. For his less talented colleagues, he programs the numbers into computers that spit out cheat sheets. The sheets outline how to react in markets where the price of an option can double or triple in minutes. All day the O'Connor traders down in the pits keep consulting their cheat sheets. They also keep pumping information upstairs, where it gets massaged into

O'Connor's computers and piped back downstairs in the form of mathematically intricate moves for playing the markets against one another. Fisher Black developed a service for supplying his customers with cheat sheets on a monthly basis. O'Connor was the first to produce them daily. Now it produces them in real time.

When the Chicago Board Options Exchange first opened in 1973, Struve's father, a businessman, gave his son a sheaf of data from the new market; he thought Clay might be interested in moving from football betting pools to stock options. Struve amused himself computing probability curves for the new market. As even a fifteen-year-old kid could see, option pricing back then was so sloppy that one could make a lot of money out of butterflies, spreads, and other simple moves in statistical arbitrage. While he was applying the necessary differential equations for figuring out what options are worth, most of the traders down on LaSalle Street were betting by the seat of their pants. They had no idea what options were worth, and their ignorance persisted for a surprisingly long time. In the late 1980s, for example, when options trading began in London, British banks arbitrarily priced every instrument at five dollars. Some were worth sixty cents. Some were worth a hundred dollars, and a lot of traders went bankrupt for not knowing the difference.

O'Connor & Associates was founded in 1977 by a mathematician named Michael Greenbaum. The company got its name from Edmund and William O'Connor. The O'Connor brothers, natives of Chicago's hardscrabble West side, made a fortune as Chicago grain traders. Bill O'Connor served as chairman of the Chicago Board of Trade, and Ed O'Connor was the main force behind the CBOT's expanding into options. As it grew bigger, the options exchange needed a clearinghouse for settling accounts, so the O'Connors started First Options. They brought in Greenbaum as risk manager for the company; later they staked his capital when Greenbaum wanted to start trading on his own.

From the moment they opened their doors as O'Connor & Associates, Greenbaum and his partners made money hand over fist. They made so much money that they were afraid other people would find out how they did it. O'Connor developed a cult of secrecy. When they bought two hundred Symbolics computers, they shredded the packing boxes to keep competitors from learning what machines they were using. Norman didn't know it at the time, but he was among the first outsiders ever allowed to walk on O'Connor's trading floor.

As more people learned how to price options, O'Connor got smart in other ways. They put together portfolios of stocks balanced by options, with the risk in one market zeroing out the risk in the other. Next, O'Connor learned that if you want to hedge two thousand stocks, you don't have to buy two thousand options. You can aggregate the risks in a single portfolio, where they might offset one another in such a way that you don't have to buy any options at all, in which case you just saved yourself the transaction costs on a lot of unnecessary contracts. This is called statistical arbitrage, and no one in the world is more mathematically astute at playing the game than O'Connor.

Struve is generally speechless, with nothing moving across his rosy face save for the hint of a smile that might come after mentally calculating a difficult equation. But at the end of Norman's presentation, he delivers some remarks that his partners interpret as wild enthusiasm. "We believe there are edges that can be collected because the market is not quite efficient," he comments. "You have to work hard and do your homework and be in the right place at the right time. There's no magic system. All the technical stuff about moving averages and trend lines is a lot of garbage. But there are statistical anomalies that are evident in a broad band of data. The anomalies are hard to find and take advantage of, but with the right microscope and the right hedging you might be able to do it. You people are on the right track. It looks good."

"This is a religious issue," adds Weinberger. "Either you believe there is structure in the markets, or you don't." By *structure* he means patterns, and patterns are predictable. The traders know the game has "edges," or inefficiencies, that could give them a statistical advantage over the house. But no one has ever discovered a way to systematize this advantage and play it in the pits.

"It takes two things to make a good trader," Struve advises Norman. "You have to understand the mathematics, and you need street smarts. You don't want to be the guy with thick glasses who is reading the sheet just when the freight train is about to roll over you. The street-smart guy will pick up a couple of quarters and get out of the way."

Everyone in the room knows that Prediction Company is the guy with thick glasses. They have zero street smarts. But Prediction Company understands the mathematics, and if someone could put the professor together with the trader, this really could produce the next Black-Scholes model. Struve wants to do a deal. Weinberger, Heimark, Solo, and the other partners want to do a deal. Jim McGill is instructed to negotiate a contract. He can't believe the terms are so favorable. All the money is there, in exactly the kind of hands-off arrangement Prediction Company has been searching for.

"I'm getting a little nervous," he says, reporting back to Santa Fe. "It's like meeting the perfect girlfriend. I've done lots of deals, but I want this one so much it hurts. This is such a sweet deal, I can't tell you how beautiful it is, and it's sweet for both sides." McGill is sent back to haggle over the final details in the prenuptial agreement, but it looks as if any day now Prediction Company will be getting married.

Walk to the Swamp

I am thinking about something much more important than
bombs. I am thinking about computers.
— JOHN VON NEUMANN

God may be subtle, but he is not malicious.
— ALBERT EINSTEIN

Before flying to Santa Fe on his emergency visit to meet the princi-
pals in Prediction Company, David Weinberger sits down to read *The
Eudaemonic Pie*. He has owned the book since it was published in
1985. "I had skipped through it, but this time I had something else in
mind," he says. "I wanted to get a feel for Doyne and Norman as people. Finance is not the right world for pure scientists. You have to get
your hands dirty, or you'll never make the transition from observer to
participant. The point of that first meeting in Santa Fe was to get a
sense for their weltanschauung, their worldview. The book gave me
a lot of confidence that these guys were willing to roll up their sleeves
and get their hands dirty."

The book recounts how a large group of intellectual riffraff—
including the author of this second volume—spent five years trying to
beat the game of roulette with toe-operated computers. A roulette wheel is a finely tooled machine for creating randomness, but
inside this randomness lies order. The game consists of a spin-

ning disc, or rotor, containing thirty-eight numbered pockets. Above the rotor is a sloping track. The croupier sets a white ball spinning on this track. About twenty seconds after its launch, the ball slows sufficiently to drop into one of the numbered cups circling below. The game re-creates a perfect universe in which two heavenly bodies dance around each other, before one of them slips out of orbit and arcs down for a crash landing. Since roulette exemplifies Newton's laws of celestial mechanics, clocking its motion should be no more difficult than landing a spaceship on Mars.

When he first hit on the idea of beating roulette, during a summer vacation spent card counting in Las Vegas, Norman saw immediately which physical inputs were required for predicting the game. Clock the ball spinning on its track. Clock the rotor spinning below. Compute their relative positions. Figure out their rates of deceleration and the arc through which the ball will travel before coming to rest in a numbered cup. The twenty seconds that exist between the start of the game and its end provide ample time for making these calculations and placing a winning bet. All he needed to do was build a computer capable of peering into the future and playing the game before the game played itself.

Norman and Doyne and a dozen fellow travelers, many of whom lived together in the big house at 707 Riverside Street, spent five years, from 1976 to 1981, perfecting the mathematical equations and building the miniaturized computers required for beating roulette. They shrank their system into toe-operated, radio-linked roulette shoes and began leading a double life yo-yoing between graduate school and Las Vegas. By the end of their last assault on the pleasure palaces of Nevada in the fall of 1981, they had acquired fame, if not fortune, for being the world's first players to beat roulette. Why they didn't acquire both fame *and* fortune stems from the difficulties entailed in shrinking computers into a shoe—a project that was undertaken before Apple Computer had built their first PC. The kinks could have been worked out with another five years' effort, but in the meantime Doyne and Norman had stumbled onto more intriguing problems. Already expert at predicting behavior in one chaotic realm, they turned to exploring chaos itself.

In 1977 Doyne Farmer, who was ditching a promising career in astrophysics, Rob Shaw, who was abandoning a nearly completed thesis on another subject, Norman Packard, who had recently started graduate school, and James Crutchfield, who was still an undergraduate, formed the Dynamical Systems Collective, also known as the Chaos Cabal. Chaos would become a growth industry in physics and a pop-cultural phenomenon, but in the late 1970s it was a shady enterprise, halfway between computer hacking and philosophy. No one was qualified to advise them, so the collective plugged away on its own. Fifteen years later the situation had changed so radically that a "chaotician" would emerge as the hero of Steven Spielberg's *Jurassic Park*. (Jeff Goldblum, who played Dr. Ian Malcolm in both the original film and its sequel, phoned Doyne to chat about how to get in character for the part.)

The Dynamical Systems Collective is best known for their work on probing chaotic systems for signs of order. Their canonical paper, "Geometry from a Time Series," published in 1980, demonstrates how one can take a dynamical system, such as gas flowing through a wind tunnel, insert a probe into the gas to measure its behavior at one particular point in time, and reconstruct from this single probe a state space picture that reveals key information about the behavior of the entire system. The collective developed a kit bag of methods for drawing these state space pictures. Their visual tricks—programmed on analog computers and digital computers and hybrid systems that patch the two together—form the basis for the techniques later applied to reconstructing the dynamics of financial markets, this time using price movements as probes.

Apart from the methods they developed for peering into turbulence and extracting the telltale traces of order, the Chaos Cabal is famous for other, less tangible, benefits to the field. They advocated science based on synthesis, rather than reduction. They hacked their way across traditional disciplinary boundaries. They borrowed liberally from information theory and evolutionary biology. They reintroduced into physics big questions about determinism and free will, and they speculated broadly on the nature of intelligence, both human and artificial. In their hands, chaos theory itself became an intriguing conundrum. It shows how randomness emerges from order, which makes long-term prediction impossible, but, at the same time, it shows how randomness possesses its own underlying order. "Randomness is in the eye of the beholder," says Doyne, "and the great promise of chaos lies in the hope that randomness might become predictable."

The Chaos Cabal were pioneers in turning computers into laboratories. They salvaged junked machines from the basement of the UC Santa Cruz physics department or built their own computers from scratch out of mail-order parts. No phenomenon in the natural world was too humble to be converted into a computer model and explored for universal properties. The best-known example of this is Rob Shaw's research on dripping faucets as chaotic systems. "We were doing garage-band science," says Doyne. "In the sixties everyone could afford to buy an electric guitar and start a rock band in their garage. The availability of new instruments caused an explosion in rock music. The same thing happened in science when everyone got hold of a computer. There was an explosion of new research as people learned how to simulate adaptive systems. Suddenly, models of very simple things began to reveal the universal properties found in more complicated systems."

Forced to disband, the Cabalists land jobs at the world's preeminent research centers, which are just beginning to get interested in the kind of garage-band science being done in Santa Cruz. The first to leave California is Doyne. In 1981 he packs his belongings into the Blue Bus and heads for New Mexico. He is going to work at Los Alamos National Laboratory's Center for Nonlinear Studies. The Blue Bus would be clocking four hundred thousand miles, if the odometer weren't broken. The bus has a jerry-built kitchen sprouting off the rear end, a leaky roof, and piebald tires. Doyne burns out the differential in the middle of the Painted Desert in northern Arizona. He finds a junkyard replacement and spends two days installing it.

When he finally chugs into Los Alamos, which is built on the edge of an old volcano overlooking the Rio Grande valley, he finds a brilliant array of scientists living in a peculiar artifact left over from the Second World War. The laboratory consists of a ragtag collection of Quonset huts and prefabricated sheds surrounded by military watchtowers, which were erected to guard the Manhattan Project, America's top-secret program to build the atomic bomb. The town itself looks like the stage set for *Leave It to Beaver*. Its suburban ranch houses are designed to highlight the city's two major domestic dramas: mowing the lawn and washing the family car.

Doyne flees back down the mountain. He moves into a small, one-bedroom cottage in Arroyo Jacona, on the banks of the Pojoaque River. He is depressed about living alone. He and Letty have temporarily split up. She is willing to abandon her career as a public interest lawyer in San Francisco and move to New Mexico, but Doyne needs another year of howling in the wilderness of male desire before he can commit to getting married. He does most of his howling a mile down the road at the Line Camp, a raucous bar that overflows on weekends with people rocking out to singers like Bonnie Raitt and John Lee Hooker.

Los Alamos welcomes Doyne with open arms, and it proves to be an exciting place to work. He has been hired to do "chaos," which means anything he imagines. All the doors are open, and the Theoretical Division is replete with experts on everything from number theory and astrophysics to molecular biology and the chemistry of uranium compounds. As an added benefit, they like to work across disciplinary boundaries.

Doyne is arriving at Los Alamos just as Mitchell Feigenbaum, the discoverer of universal scaling laws in the transition to chaos, is leaving. So the new postdoc is encouraged to fill the vacuum created by Feigenbaum's absence. Doyne invites scores of visitors, organizes conferences, and recruits new postdoctoral fellows in nonlinear dynamics, many of whom go on to become luminaries in the field. Doyne himself becomes an Oppenheimer Fellow and staff member in the mathematical modeling group, before being asked to found a new group in complex systems—the first of its kind.

The intellectually exciting climate comes with a few minor irritations. Less than half the research at Los Alamos is open to public scrutiny. The rest takes place "behind the fence," in areas that only people with security clearances can enter. This "fence" sometimes divides corridors in half and necessitates long walks for uncleared people who are trying to move around the lab. Doyne has no interest in working on military secrets. He attaches to his published papers a request that readers "use these results for peaceful purposes." But the computer facilities at Los Alamos are behind the fence, and he has no way to use them without passing a security check. It takes him fifteen months to get cleared.

In the meantime, Doyne convinces the Center for Nonlinear Studies that microcomputers are the wave of the future. A new company called Sun Microsystems is selling what they call "scientific workstations." When Doyne calls them up, Sun is so excited about the possibility of selling a machine to Los Alamos that Bill Joy and John Gage, two founders of the company, fly out to New Mexico. They end up shipping Doyne one of the first workstations off the assembly line, serial number 0042, running version 0.9 of the operating system. Not surprisingly, the machine has a few bugs, and Doyne spends a lot of time making repairs down in the bowels of the operating system.

He scrounges analog computers from the Los Alamos salvage yard and buys more equipment on government grants. Soon he, Norman, and the other members of the Chaos Cabal, who have become regular visitors at the lab, are outfitted with enough electronic gear to form their own research center. Unfortunately, this motley collection of hardware attracts the attention of the people at Los Alamos in charge of health and safety. While Doyne makes his rapid ascent up the bureaucratic ladder to Group Leader, he is also being cited for workplace violations. To avoid the noise from his computers, he stores his disk drives in the hallway outside his office. These machines are not secured against enemy intruders. His filing cabinets are improperly stacked on top of one another. He has too many books on his bookshelves, which could prove dangerous in the event of an earthquake. His wiring is faulty, his plugs are improperly grounded, and so on.

Asked to provide the entertainment for a Theoretical Division Christmas party, Doyne pokes some good-natured fun at the Energy

Department by writing and directing a farce called "The Ghost of Oppenheimer Past." The dark forces at Los Alamos are represented by "Dr. Edward Tattler," who may or may not resemble the inventor of the hydrogen bomb. The forces of light are represented by Robert Oppenheimer, the physicist who founded Los Alamos and who later directed the Institute for Advanced Studies, where he barely survived one of Joe McCarthy's anti-Communist witch-hunts.

Doyne's play opens with "Dr. Eddy," played by Chris Langton, careening around the stage in a wheelchair. He is busy throwing pieces of raw liver into a beaker, trying to bring an alchemical experiment to life, when he decides the missing ingredient is cold fusion. He commands his assistant to find something cold. Into the pot goes a block of dry ice, which creates a lot of bubbling and smoke. The lights go out and a string of firecrackers explodes onstage.

The narrator tells us that Dr. Eddy has just "given birth to an entirely new concept in our struggle for dominance over the communist horde." He screams out the name of his great discovery: "COLD war conFUSION."

In Scene Two, a new postdoctoral fellow arrives at Los Alamos. The lab is so overcrowded that his office turns out to be a toilet in the Theoretical Division. The postdoc begins filling out the endless security forms that are required by the Energy Department in Washington. The director of the lab, played by Doyne, explains to the postdoc that security is the primary concern at Los Alamos, and thanks to COLD war conFUSION, the lab has discovered a brilliant way to protect America's scientific secrets. It has adopted as its motto the expression, "JUST SAY NO TO SCIENCE." This slogan is disseminated on posters showing the head of Albert Einstein surrounded by a circle with a slash through it.

"We've completely outwitted them," proclaims the director. "With no science, there are no secrets to steal."

The postdoc, sitting on his toilet seat, has allowed his mind to wander. He is mumbling to himself and jotting down the numbers in an equation, when he is caught in the act by a jackbooted member of the "GAOstapo." (The Government Accounting Office, or GAO, is the watchdog agency that necessitates much of Los Alamos's paperwork.)

The security guard warns the postdoc that he risks a jail sentence for not filling out his "minute-by-minute effort report."

The postdoc is dragged off to Washington to be interrogated by the secretary of energy. He receives "the maximum penalty," which is promotion "to a high-level position as an administrator." Suddenly the windows and walls in the secretary's office begin to rattle. The lights flicker, and smoke furls under the door as the Ghost of Oppenheimer floats into the room. The ghost effects a reverse time machine. He takes the secretary of energy back to the 1940s to show him what Los Alamos would have looked like if Tattler and friends had been running the laboratory during the war.

The stage goes dark. Up go the lights, revealing Oppenheimer, Enrico Fermi, the postdoc, the GAO agent, and the secretary of energy clustered around the first atomic bomb. "Well, Enrico, it looks like we're finally going to do the big test tomorrow," announces Oppenheimer.

"Can you pass me that wrench," says Fermi, in a thick Italian accent. Before Oppenheimer can hand him the wrench, the GAO agent grabs it out of Oppenheimer's hand and demands that he fill out a work order.

"But, we're gonna testa the bomba tomorra," protests Fermi.

The agent begins interrogating him to see if Fermi has the appropriate security clearance. By the time Fermi has filled out the correct forms, the war has been over for thirty years.

The secretary of energy is shaken by Oppenheimer's revelation. He suddenly realizes that Cold war conFUSION "is . . . a communist plot." He suppresses the security forms and adopts as the laboratory's new motto, "JUST SAY NO TO BULLSHIT."

Norman, like Doyne, leaves Santa Cruz for an intellectually stimulating environment, although one less quick to embrace the new world of computers. In 1982 he moves to France to begin a NATO postdoctoral fellowship at the Institut des Hautes Etudes Scientifique. Located outside Paris, IHES is the French equivalent to the Institute for Advanced Studies. But instead of a technologically advanced research center, Norman finds a sleepy enclave of theoreticians whose computational environment consists of one Hewlett-Packard calculator that they share among themselves. He arranges to use the computer at a nearby technical school. It turns out to be a mainframe computer programmed with batches of cards. Finally, all Norman can do is take out his pencil and start theorizing along with the rest of them.

Things begin to look up the following year when he moves to the Institute for Advanced Studies itself. Founded by a retailer who had the good fortune to sell out to Macy's six weeks before the 1929 stock market crash, the institute opened in 1933 with the arrival of Albert Einstein, its first professor. Einstein was so famous by then that his appearance in New Jersey, as one observer remarked, was equivalent to having the Pope move to a church in the Pine Barrens. By the time it was sheltering Kurt Gödel, John von Neumann, J. Robert Oppenheimer, Freeman Dyson, and other great names in twentieth-century science, the institute had become as well known as its neighbor, Princeton University, with whom it shares a quaint colonial town, but no other affiliation.

Built as "a paradise for scholars," the institute coddles its scientists so well that they have not a care in the world other than devoting themselves to pure research, and nothing but pure research. They are forbidden from dabbling in practical matters. Von Neumann, for example, got in trouble for building one of the world's first computers. After his death, the machine was removed to the Smithsonian Institution and the administration adopted an ironclad rule against scientists being allowed to tinker in their laboratories. Thirty years later the purists are beginning to equivocate. Norman is hired as part of an effort to bring computers back into the institute. He moves a raft of machines into his office, and what are computers in the hands of the Chaos Cabal other than desktop laboratories?

Norman and fellow cabalist Rob Shaw are working with Stephen Wolfram, who is charged with starting a complex systems research group in Princeton. Wolfram at the time was a twenty-four-year-old British prodigy. He had published his first scientific paper at the age of fifteen. Five years later he had completed a Ph.D. in particle physics at Cal Tech and was about to become the youngest recipient of a MacArthur Foundation "genius" award. Pudgy, balding, with a dark complexion and outsized ego, Wolfram bears a striking resemblance to Napoleon. But rather than merely conquering Europe, Wolfram intends to master the terra incognita of complexity.

Wolfram's research group is installed in an airy, skylit suite of rooms on the third floor of Fuld Hall, two floors above Albert Einstein's old office. Fuld Hall is a big Georgian-style structure that sits in a square mile of manicured lawns and woodlands, which are dotted with lakes and a marsh that doubles as a bird sanctuary. The institute is a self-contained world of its own, complete with library, residential apartments, dining room, and living room, in which tea and cookies are served every afternoon. Wolfram buys the first Sun workstation to appear at the institute, and soon his third-floor suite is packed with computers and other scientific gear, including the B. C. Wills roulette wheel that Doyne and Norman use for their gambling research.

Leaving behind his dripping faucet, Rob Shaw begins working with Norman on something called spatial dynamics, which encompasses any system that changes dynamically in time, like the surface of the Earth. These systems are studied by means of lattice map models. These are the simplest versions of chaos, little dynamical systems in which one variable interacts with itself to generate random-looking orbits. Things begin to get interesting when out of these apparently random orbits peak the owlish masks that belie order in chaos. Wolfram himself begins studying another form of spatial dynamics known as cellular automata. Pressing his campaign with didactic fervor, he is convinced that all the major unanswered questions in science, ranging from the origins of life to the laws of complexity, can be solved through studying cellular automata.

Cellular automata are another legacy of von Neumann's that the institute has left unexplored since his death. They are a beautiful example of how complexity can arise out of underlying simplicity. They consist of nothing more than grids of cells that blink on and off, changing state according to a few simple rules. One of the better-known examples of a cellular automaton, The Game of Life, was invented in 1970 by Cambridge mathematician John Conway. Life involves the manipulation of tokens on an expanse of graph paper, or, as it is played today, inside a computer. These simple systems can generate a surprising range of complex patterns, such as waves and other kinds of motion observed in nature. This leads Wolfram to speculate that nature itself might be a cellular automaton; the complex patterns we observe around us have arisen from nothing more complicated than the binary flip-flops of cellular automata in action.

What intrigues researchers about cellular automata is their ability to simulate galaxy formation and other phenomena that appear to contravene the second law of thermodynamics. Instead of being worlds that run down from order to entropy, cellular automata, through interaction with their environments, actually evolve in the opposite direction, from randomness to order. Norman has produced a striking example of this behavior by taking a grid with a single colored seed and digitally evolving it through enough generations to produce an artificial snowflake. One of the ultimate goals of this research is to discover the general laws of "self-organization."

Cellular automata provide a mathematical formalism for studying biological processes such as self-reproduction, the immune system, evolution, and even the origins of life itself. Wolfram and his group in Princeton, and Doyne and another group of researchers at Los Alamos, are soon tackling these questions head-on. At a big Los Alamos conference on cellular automata organized by Doyne and Wolfram in 1983, Doyne addresses the lifelike features of cellular automata by describing what he calls the second law of self-organization. There is no first law, and the rest of the idea is still being worked out, but the second law might be defined as follows: *Matter tends to organize itself*.

Norman and Doyne, coauthoring many of their papers, will spend the next decade trying to define the basic terms in a theory of self-organization. This work is geared toward understanding how order emerges at the edge of chaos. *The edge of chaos* is a phrase coined by Norman. It first appears in a paper he publishes in 1988 called "Adaptation Toward the Edge of Chaos." The phrase is now one of the lodestar concepts in the theory of complex adaptive systems. It refers to the hypothesis that the boundary area between chaos and order is the most fruitful place for complex systems to emerge. This is where "noise" turns into "information" and the self-replicating patterns of life, or artificial life, as the case may be, begin to appear. Cells, brains, organisms, ecosystems, corporations, and economies are all thought to exist at the edge of chaos, in a regime not so ordered as to be sterile and not so chaotic as to be meaningless. Life has to "steer a delicate course between too much order and too much chaos, the Scylla and Charybdis of dynamical systems," says Chris Langton, the scientist who invented the term *artificial life*, before going to work with Doyne at Los Alamos.

Pushing ever deeper into using computers as living laboratories, Norman and Doyne team up with biologist Stuart Kauffman. The problem they tackle is the origin of life. How does nature develop from prebiotic soup to primitive metabolisms and on from there to living organisms? They publish a series of papers outlining a possible evolutionary pathway. They devise models for studying the chemistry of molecules. They figure out the chemical networks by which the first metabolisms might have emerged.

Along with inquiring into evolution at the molecular level, Norman and Doyne, working with biologist Alan Perelson, also begin researching the immune system. This mechanism, by which the human body distinguishes "self" from "other," opens a rich vein of inquiry into pattern recognition, memory, and learning. These are the themes of another big Los Alamos conference organized by Norman, Doyne, and other lab scientists in 1985. The conference is called "Evolution, Games, and Learning: Models for Adaptation in Machines and Nature." Computers by then are being used to simulate the adaptive behavior known as evolution, and they are beginning to display what look like the first faint stirrings of their own adaptive behavior. Computer models of ants, amino acids, and neurons show how collective behavior is qualitatively different from the sum of its individual parts. The idea of machine "learning" is still a high-blown metaphor, but thinking machines are getting smart enough to provoke speculations about the onset of "artificial life." This is the title for another big Los

Alamos conference organized in 1987. The conference is followed by a second A-life gathering in 1990 entitled "Emergence and Evolution of Lifelike Forms in Human-made Environments."

The philosophical implications of this research are summarized in a paper that Doyne and Letty authored in 1989. The paper is written for a conference honoring the physicist Murray Gell-Mann on his sixtieth birthday. The event is organized around sixteen questions that Gell-Mann, a Nobel laureate in particle physics, and one of the founders of the Santa Fe Institute, considers the greatest unsolved challenges in science and human affairs. Doyne and Letty address these questions in a paper called "Artificial Life: The Coming Evolution."

Carbon-based life on earth, which developed "under the control of protein- and DNA-templating machinery," is about to witness the emergence of new human-made systems that are capable of reproducing themselves. Computer viruses are one example. Self-reproducing machines and cellular automata are other examples of artificial life. Doyne and Letty speculate that a new class of organisms, originally designed by humans but eventually autonomous, will assume an evolutionary life of their own within the next fifty to a hundred years. "The advent of artificial life will be the most significant historical event since the emergence of human beings," they write.

It is their use of computers as laboratories and their speculations about artificial life and the evolution of complex systems that lead Norman and Doyne to think about money. The fateful moment occurs during a walk to the Princeton marsh in the spring of 1985. They have been up all night, programming like crazy, working on their origin of life model. At dawn, they stumble out of Fuld Hall onto the lawn where Einstein used to take his daily constitutional. It is a crisp morning, with buttery light filtering through the trees. In a daze, they walk across the lawn to the lake. They follow the path into the woods and on from there to the marsh, which is full of birds singing in the morning light.

Norman, who lives in an apartment facing the library, often works into the early hours of the morning. Then he strolls home, rolls himself a cigarette, and listens to the late Beethoven quartets or Monteverdi before falling asleep. He gets out of bed in time for lunch, which is provided by the institute's Hungarian chef. This is followed by afternoon tea in the living room and dinner in the dining room. The institute actually has two afternoon teas, carefully arranged so that the natural scientists can avoid talking to anyone but themselves. When Norman, a physicist, begins attending math seminars, this is the most interdisciplinary activity the institute has seen in decades. Some of the guests he invites for tea are also considered unusual.

The day before their walk to the marsh, Norman and Doyne had been on a similar teatime stroll with Stephen Wolfram and a visitor to their laboratory, Aristid Lindenmayer, the Hungarian botanist who invented something called L-systems. These mathematical constructs for describing embryonic growth and the development of plants are a kind of cellular automaton, and Lindenmayer, before his death in 1989, was a pioneer in the field of complex systems research, which he began studying long before the term existed.

Traveling with Lindenmayer is his son-in-law, who is the founder of a company that makes robot arms. He has just sold his company for a large sum of money and is looking for ways to invest it. He wants to apply what he knows about robot arms to building a system for predicting the stock market. "Your work on machine learning and chaos might be useful in some commercial way," he suggests to Norman and Doyne. "The stock market might be a chaotic system good for modeling. Have you thought about using your methods for financial forecasting?"

Doyne and Norman return to this question the following morning on their walk. Will machine learning help build a system for stock market prediction? Is nonlinear forecasting potentially useful for making money? They decide the answer is *yes*.

The Ghost of Oppenheimer

What we see around us is surface complexity arising out of deep simplicity.

— MURRAY GELL-MANN

The people who master the new science of Complexity will become the economic, cultural, and political superpowers of the next century.

— HEINZ PAGELS

A few weeks later, in the spring of 1985, Norman and Doyne meet again in New Mexico. They spend the day working in Doyne's office. At nightfall, they drive off the Los Alamos caldera. The top is down on Doyne's sports car. The radio is cranked. Across the Rio Grande valley rise the muscular peaks of the Sangre de Cristo. The snow-capped mountains are torched red by the setting sun. The valley stretches below like a deep blue sea. The land is furrowed with dry wash rivers and canyons and mesas melting into the aqueous light. The road passes through alternating layers of umber and mustard stone as it descends through the geological record, backward through time. Los Alamos disappears behind them. Soon, the apex of western science, where the sun's energy was first harnessed into a nuclear reaction, dissolves into nothing more than a speck on the side of an ancient volcano.

"I have a new idea about how to use low-dimensional chaos for forecasting," Doyne shouts over the engine noise and wind. "It's really pretty simple. You embed the time series in a state space and fit functions to the data. To deal with nonlinearities, you fit different functions in different places. To make your forecasts, you just find the right function and apply it."

Norman immediately gets the point. "You could use this for anything from sunspots to the stock market," he says.

They begin an animated discussion about some of the basic ideas that will later be used at Prediction Company. The technique involves fitting functions to time series. Imagine a stream of data, a record of daily sunspot activity, for example. The sunspot activity on two successive days is two numbers, which can be represented as points in a plane. The sunspot activity on the *next* day can be thought of as a point suspended above this plane. The picture of sunspot activity for a year will be a cloud of points in three dimensional space. To make a forecasting model, one fits a surface, like a rumpled sheet, through this cloud of data. If the sheet fits in close proximity to all the points, it will be a good model. Forecasts are made by stretching the sheet into regions where additional data points will be recorded when the future rolls into the present.

"Do you want to collaborate on the project?" Doyne asks Norman.

"I'll pass," he says, staring across the valley at the last light coming off the mountains.

Doyne and Norman are parting ways, with Doyne moving into forecasting and Norman sticking with snowflakes and other exotic dynamical systems. The irony is that five years later, traveling by different routes, they will have arrived at exactly the same place. Norman has two reasons for declining Doyne's invitation. "I wasn't keen on learning all the classical statistics you need to know about model building in order to embark on this big program," he recalls. There is a world of difference between isolating patterns in complex data and building systems to trade on these patterns. This requires a headlong dive into statistical analysis and testing. Is one really seeing patterns, or phantasms? Do they endure, or disappear? Norman at the time thought statistics was boring, or at least not as interesting as growing synthetic snowflakes. He has subsequently changed his mind.

He now thinks statistics is a deep philosophical issue, and he delights in bringing rigor into financial forecasting, which until recently has existed in a statistical Stone Age.

"I'm more interested in inventing something out of the blue," he says. "I've tended scientifically to fire potshots off into the distance. Sometimes I hit something. Sometimes I fall in the mud. It's not my style to start with a firm body of knowledge and push it out incrementally, which is what Doyne was proposing. Doyne and I work together well because of this complementary approach to scientific problems. So later, when we got back together on this idea, there was a nice fit."

Norman has a second reason for declining Doyne's invitation. "It wasn't the healthiest thing for us, intellectually or professionally, that we were always connected in our research. I was feeling pressure to establish myself independently." Many of Norman's scientific papers are coauthored with Doyne or other members of the Dynamical Systems Collective. But this practice is discounted in academia, where only the lead author on coauthored papers is given credit for the work.

The idea of using chaos theory as a predictive tool has long intrigued the four members of the Chaos Cabal. Even back in Santa Cruz in the 1970s the group thought about applying their research to financial data. They played a game during their meetings at Café Pergolesi. They asked themselves, what nearby phenomena look random but actually contain hidden structure? The smoke curling up from an ashtray? The vortex formed around a sugar cube dropped in a cup of coffee? The stock data listed in the newspaper lying on the table? The idea kept cropping up, especially when they were discussing the dearth of grants available for scientific research, that beating the world financial markets might be a good source of funds.

"Any time you're analyzing a complex, random-looking signal and searching for hidden structure, it's natural to think of financial data," Norman explains. "The only difference between financial time series and other time series is that they present problems of interest to people with money."

In 1985 Doyne gets a grant from the U.S. Air Force to research nonlinear forecasting. The grant gives him enough money to bring his friends to Los Alamos for the summer, although he has no place to put them, the lab being full. So he drives down the mountain to the neighboring town of El Rancho and rents the former general store and speakeasy. This is a large adobe building with fifteen-foot, pressed-tin ceilings, a full kitchen, and enough bedrooms to double as a hotel.

Rob Shaw arrives in the Cream Dream, his 1959 Ford station wagon, which is packed with computers. Norman hauls a couple of workstations out from Princeton. Jim Crutchfield borrows a machine from the University of California at Berkeley, where he is working as a postdoc. Doyne drags his gear down from Los Alamos. They rewire the building with an ethernet. They install a Ping-Pong table and open themselves up for business as the El Rancho Institute.

Their normal workday begins sometime in the afternoon. They take refuge from the desert sun behind their adobe walls, until day shades into evening, when they venture outside for cocktail hour and a barbecue dinner. Then they jam all night, except for a break from midnight to 2:00 A.M. for the daily Ping-Pong tournament. Norman is the best singles player. Doyne, with a strong forehand, but no backhand to speak of, is best in doubles. Rob is quick, although he doesn't make much use of spin. Jim is the most aggressive. As the youngest member of the Chaos Cabal, joining while he was still an undergraduate, he has always been a bit scrappy in defending his corner. Rounding out the player list are Chris Langton, Emily Stone, and another half dozen El Rancho researchers. The only person sitting out the tournament is Stephen Wolfram, who keeps working no matter how lively the action gets around him.

On the menu of projects to be explored that summer are genetic algorithms, nonlinear forecasting, cellular automata, dynamical systems, the origins of life, and . . . roulette. It was roulette that inspired Doyne's foray into forecasting, and he decides to revisit his original effort at predicting the future. He will improve the roulette computer by reprogramming it with new software. The original, Newtonian equations will be replaced by the kind of empirical time series techniques that he and Norman are using to analyze chaotic systems. He has yet to realize it, but this will be his first step in moving from the gambling casinos of Las Vegas to the bigger casino known as the world financial markets.

Prime researcher on the roulette project is Stephen Pope, a twenty-three-year-old back-country ski guide and itinerant physicist. A compact, speedy young man with blue eyes and a shock of brown hair, he is hyper-talented in several domains. He is an accomplished jazz pianist. He wields a fierce Ping-Pong paddle, and he knows computers inside and out. As an undergraduate physics major at UC Santa Cruz, he worked with Rob Shaw on dripping faucets as chaotic systems. After graduating in 1984, he was driving a truck cross-country, when he stopped to visit Los Alamos. Doyne had been trying for several weeks to patch together a system for getting his analog and digital computers to talk to each other. Pope looked at the wiring diagram and saw at a glance the one missing connection. Doyne offered him a job on the spot. Pope declined, preferring instead to spend the winter as a ski bum in California. But he returns to New Mexico the following summer, where he bunks in one of El Rancho's empty rooms and goes to work rewriting the roulette program.

The introduction of nonlinear dynamics into roulette prediction is linked with a big push to develop fail-safe hardware. The project has been plagued for years with short circuits in bra-mounted battery packs, shocking shoes, bug-ridden computers, and other electrical glitches that kept team members rushing to the toilets for ad hoc repairs. In 1983 Doyne had placed an advertisement in *Gambling Times* looking for investors to build a new, improved roulette system. Among the respondents was Fred Britton, a Canadian inventor, who offered to assemble a team of gamblers and play the Eudaemonic system on a franchise basis. His team would finance building the next generation of roulette shoes. Then they would trot around the world knocking off casinos, while depositing the occasional royalty check back in the Eudaemonic Pie.

When not speculating about deterministic chaos and roulette, the El Rancho researchers are working on other complex problems. Stephen Wolfram and Rob Shaw are pushing deeper into cellular automata. Jim Crutchfield is linking information theory to chaos. Norman Packard is evolving synthetic snowflakes. And Doyne Farmer is embarked on his forecasting project. Over the next few years, he will author a flurry of papers on the subject, including a major work called "Predicting Chaotic Time Series," coauthored with John "Sid" Sidorowich and published in 1987.

Their recipe for extracting useful information from dynamical systems—including stock markets—begins with embedding data from these systems in state space. A dynamical system consists of two parts. The essential information about the system, called a *state*, and equations of motion that describe how this state evolves through time. An evolving system is represented as an orbit in state space. This is a multidimensional graph whose coordinates are the components of the state. In a mechanical system, like a game of roulette, these coordinates will be position and velocity. In an ecological model, they might be the population size of different species. In the stock market, they will be things such as opening and closing prices.

Equations of motion provide explicit rules for generating orbits from these coordinates. These orbits can be launched from a single point in state space, and they will change in time as the variables themselves are altered. State space pictures are useful for representing behavior in geometric form, and it is from the geometry of the time series that one then discovers the order in chaos. It is these pictures of data points in motion that allow one to distinguish between deterministic chaos and plain old randomness.

In 1986 the "dynamical systems mafia" at the Institute for Advanced Studies pack up their computers and move away from Princeton. Stephen Wolfram's relationship with the institute has always been a bit strained. They don't like his tinkering with computers and his research is too worldly, if not downright commercial. Relations get frosty when Wolfram begins helping his friend Danny Hillis program the massively parallel computers he is building at Thinking Machines. They become positively glacial when Wolfram begins selling his own high-powered computer program called Mathematica. Wolfram has already stormed out of a job at Cal Tech. This followed a bruising battle over intellectual property rights, and with similar conflicts brewing in Princeton, he decides it is time to move on.

He announces that he and his team are on the intellectual job market. The entire group can be bought as a package deal. Rather than being *at* a research institute, Wolfram intends to *become* a research institute. Among various offers, the best bid comes from the University of Illinois. The university will give everyone in the group faculty positions. They will install Wolfram and company in their own building, called the Center for Complex Systems Research, and wire this building into the university's supercomputer center, which offers one of the world's biggest silicon breeding grounds for cellular automata and other forms of artificial life.

Although the university is providing the infrastructure and putting up the initial seed money, Wolfram's institute is supposed to secure the rest of its budget from private and public donors, but asking for money has never been one of Wolfram's strong points. Soon he is so involved in developing Mathematica, his program for doing calculus and other symbolic computations, that he quits the university and goes into business as founder and president of WRI, Wolfram Research International. "Mathematica will be to complexity what Newton's Principia was to mechanics," Wolfram announces with no modesty. At the same time, Rob Shaw wins a MacArthur "genius" award and moves back to Santa Cruz. This leaves newly appointed assistant professor Norman Packard holding the bag as director and chief fund raiser of the Center for Complex Systems Research.

When not hustling for money, Norman is trying to figure out how snowflakes grow. In the process, he invents a new kind of learning algorithm that is good at predicting dendritic growth. This is the accretion of lattices—little kaleidoscopic patterns or fractals—by which snowflakes enlarge themselves from single seeds. Soon after he invents it, Norman's snowflake algorithm gets applied to an unusual set of data, where it performs so well in sifting through the numbers and predicting various outcomes that Norman realizes he has invented a general modeling tool that can be applied to any set of numbers—including financial numbers.

In the spring of 1989, with a fellowship from the Sloane Foundation, Norman and his wife, Grazia, move to Italy for six months. This is when Norman, in exchange for an office borrowed from his brother-in-law, agrees to analyze data for *la Regione Lombardia*. This governmental body directs seven hundred and thirty suboffices throughout Italy's industrial north. For several years these offices have been filling out questionnaires detailing how they operate. The questionnaires are piling up, and someone is supposed to examine these data and figure out which procedures result in the most efficient spending of public money.

When Norman plugs the Lombardy numbers into his snowflake program, he begins seeing lots of "structure," which, in this case, refers to bureaucrats who are spending public money efficiently. A genetic learning algorithm, which is modeled after biological evolution, wraps lengthy number strings around one another in the same way that human bodies are compiled from strands of DNA. The algorithm, by performing its own versions of "mutation" and "crossover," leads these number strings to "reproduce" themselves through successive generations. The algorithm plucks from each generation the most biologically "fit" patterns, which continue evolving through further mutations and crossovers. The advantage to doing evolution in a computer is that experiments take place in nanoseconds rather than millennia.

Conceived in snowflakes and brought to parturition by the Italian government, Norman's algorithm is christened with the hopeful name Prophet. Prophet looks at any kind of complex data, makes hypotheses about fruitful correlations, and then zips through the numbers testing these hypotheses. "Populations" of hypotheses are challenged by changing environments, and those that evolve successively are kept alive; the others die off. To be evolutionarily successful, a population has to do a good job of projecting past success into future performance. It is irrelevant to Prophet what kind of time series it is looking at. Plugged into the computer, this information is nothing more than data streams, bit strings, the endless natter of 0s and 1s blinking through logic gates. When it reemerges from the model, the information has been massaged into evolutionarily fit hypotheses and correlations, some of which are good enough to bet on.

The Lombardy report is a success, and Norman begins looking for financial time series to pop into his model. This is followed by a job offer from Wall Street and a phone call to Doyne alerting him to the fact that Norman is thinking about going into business.

Doyne dons his "Eat the Rich" T-shirt for the March 1991 meeting at which Prediction Company is founded. This is the watershed rendezvous at Jim Pelkey's house when the predictors decide to go into business together. Pelkey will scare up the money. McGill will handle business. Doyne will cut his ties to Los Alamos, and Norman will abandon his newly founded research center and move to Santa Fe.

In May 1991 Doyne gives notice at Los Alamos that he is quitting. The news catches his bosses by surprise. Four years earlier he had been named leader of a new complex systems research group. He had juggled budget lines with enough acumen to hire twenty people and host other visiting researchers who filled his building to overflowing. This began an incredibly productive period in Doyne's life as a physicist. While leading three major research efforts into artificial life, machine learning, and dynamical systems theory, including forecasting, he published more scientific papers than the thirty-nine years in which he had been alive. In 1990 he was awarded the prize for doing the best published research at Los Alamos.

At the time he was picking up his award, the money for doing this research was drying up. Doyne was writing nine proposals for every project that got funded, and the process was becoming increasingly politicized. The Cold War was ending, which meant that previously well-financed bomb builders had begun scrambling for the same basic research money as everyone else at Los Alamos. Doyne may have ranked among their best scientists, but he was still being hounded by the Energy Department for storing his computers in the hallway and stacking his filing cabinets too high. When fighting the budget wars got too intense, Doyne found the Ghost of Oppenheimer whispering in his ear, "Now is the time to make a break. Go for it!"

When news got out that he was leaving the lab, Doyne was contacted by a reporter from *Science* magazine. His comments were highlighted in the next edition. "Farmer's idea is to adapt the sophisticated algorithms of 'complexity theory'—an extension of chaos theory—to predicting how cash will move in financial markets. Getting rich, he says, is his last, best hope for doing more research.

" 'Things have reached this state,' he says, 'because the tradition of support for high-risk, high-payoff research at Los Alamos has virtually come to an end.'

" 'If it works, then the research will make money by itself, and I won't have to convince some bonehead back in Washington that it's worthwhile.' "

The use of the word "bonehead" generates some appreciative chuckles in the scientific community. "I seem to have struck a chord," says Doyne. At the end of the month, he packs his Los Alamos life into boxes and moves out of his office. It is almost ten years to the day since he first arrived on the hill.

Marrying for Money

The game of professional investment is intolerably boring and overexacting to anyone who is entirely exempt from the gambling instinct; whilst he who has it must pay to this propensity the appropriate toll.

—JOHN MAYNARD KEYNES

The gambling passion lurks at the bottom of every heart.

—HONORÉ DE BALZAC

It looks to be the perfect marriage. O'Connor, with its street smarts, allied to the brainy researchers at Prediction Company. Now comes the hard part. After agreeing to the arrangement, O'Connor and Prediction Company sit down to negotiate a contract. They begin with a letter of intent—a kind of prenuptial agreement—which will be followed by the legal tying of corporate knots.

On a blustery day in March 1992, negotiators for the two sides meet in a conference room on LaSalle Street. Seated at the table, with a view through the room's glass walls onto O'Connor's trading floor, are fast-talking David Weinberger and closemouthed Clay Struve. Weinberger is dressed in a sea green sports shirt, chinos, and boat shoes without socks. He is sporting a nice tan from his spring ski vacation, but a bad back keeps him hopping up from his chair to pace the room.

Blond, mustachioed, imperturbable, Struve is dressed in chinos, black loafers, and a blue sports shirt monogrammed with a little polo player. He keeps an eye cocked on the red digits flashing across

O'Connor's electronic tape. Rarely does he growl out a remark, and most of these utterances are at least twenty-five percent numerical. Even when silent, there is some aura to Struve, the old rugby pro, that makes him feel like the central player in O'Connor's scrum.

With Doyne still in Australia lecturing on chaos, Prediction Company is represented by Norman, who looks wrinkled but cosmopolitan in a blue linen sports coat and dark pants. Seated next to him is the poker-faced Jim McGill. He is dressed in the same casual clothes as his hosts, but without the polo ponies. He types notes into a portable computer as the conversation swings from money to morals. He has two truisms that summarize this perilous moment when business alliances are formed: *You can't do a deal if the chemistry isn't right,* and *All the bad news comes after the contract is signed.*

Speaking little, McGill is hewing close to another of his zen business precepts: *Never make the first move in a negotiation. Always wait for your opponent to reveal his strengths and weaknesses, before countering with your own offer.*

Unlike McGill, Weinberger is a master of jawboning. Oblivious to the zen, he is in the process of laying his cards on the table straight off. Out they come, *ping, ping, ping,* and then he gets up and starts doing stretching exercises while hanging on to the back of his chair. Prediction Company and O'Connor will sign a five-year contract. The terms of the contract, to be kept secret, can be summarized as follows. O'Connor will put up the money to fund the development of Prediction Company's trading technology. A much larger chunk of capital will be used to deploy the system.

Prediction Company will get a slice of the profits. (A figure between ten- and twenty-five percent is standard in the industry.) Out of their slice, they are expected to repay O'Connor's investment in developing the technology, but the rest of the money is theirs to keep. Until a contract is drawn up and signed, O'Connor will pay enough "good faith" money to cover Prediction Company's operating expenses. The contract will have a "no-cut" clause for the first two years. If O'Connor backs out of the deal before then, Prediction Company is not obliged to repay the money invested in it. At the end of five years, assuming their obligations are settled, either party can walk away from the deal essentially free and clear.

McGill can hardly believe what he is hearing. "They're putting up all the money, and they're not asking for an ownership stake in the company? This is the most incredible deal I've ever negotiated!"

Norman is more skeptical, and his doubts are confirmed when he phones Doyne in Australia to report on the negotiations. One of the kickers in the contract—and the reason why *O'Connor* thinks it's a sweet deal—is the termination clause. If, at the end of their five-year relationship, Prediction Company decides to do business with someone else, O'Connor will get a copy of their tools. This includes the source code for their algorithms, model-building programs, execution software, and models themselves. In other words, Prediction Company's entire bag of tricks.

"This makes me really nervous," Doyne frets on the other end of the line. "If they get our models, what do we have left? Where's our bargaining power?"

Weinberger has explained that this is required to offset the risk on his side of the table. "We think you'll be happy enough with the deal that in five years you'll want to extend the contract."

"It doesn't make sense!" Doyne says from Australia. "It's too dangerous to hand over everything we've done to someone who can kiss us good-bye and walk away with it."

"I agree with you," says Norman. "It's a bit hard to take."

Trying to keep everyone calm, McGill gets on the phone. "There is no stock," he says, reviewing the terms of the contract. "They're paying all our bills. It has no cut for two years. If they leave before then, they walk away from their money, and we still own as much of the company as we do today."

"I thought they were charging interest on that money," says Doyne.

"They are," explains McGill. "It's an advance against trading profits. But if they walk early, we don't owe it to them. It's a very sweet deal," he repeats, trying to convince his colleagues. "The reason it works, in a nutshell, is that they get a copy of the technology at the end of the deal. This is what makes the whole thing play. None of the other people we've talked to, except the ones who want to buy us, would know what to do with our technology if they got it. But O'Connor *does* know, and this gives us an extra chip to put on the table. It's what makes the valuation balance on both sides."

"Who knows where this technology is going to be five years from now," Norman suggests. He is beginning to appreciate McGill's point of view. "There are lots of unknowns in the equation, and it might be worth taking the risk."

"The deal gives us two years of focused effort on the problem, to see if it can be cracked," says McGill.

"I still don't like the idea of giving away our ideas," says Doyne. He promises to think about the deal, without rejecting it out of hand, while Norman and McGill head back to the negotiating table to iron out the details.

In spite of his volubility and manic presence during these discussions, David Weinberger does not reveal two substantive facts. He does not tell Prediction Company why he is in a hurry to do the deal. Nor does he tell them how much fun he is having. As he remarks to Struve, "They remind me of us, when we've been the little guys sitting on their side of the table."

Along with convincing Prediction Company to do a deal with O'Connor, Weinberger also has to convince his partners that they should welcome these strangers into their midst. A company that shreds computer boxes to hide its technology from competitors is about to link itself electronically to an adobe bungalow in Santa Fe.

Weinberger argues that Prediction Company and O'Connor are not really in the same line of work, that hedging bets in the derivatives markets is different from trying to predict which way the markets will move. Prediction Company is a new spin on the game, with advanced technology that O'Connor can use. He makes it a virtue that they are out in the desert, away from O'Connor's daily operations. "They won't be roaming around here making people nervous," he says.

"Prediction Company underestimated the difficulty of the problem, and they didn't know enough about markets to know what they didn't know," Weinberger concludes. "But the chemistry was good. We trusted these people at the gut level. This is crucial if you're going to put your money behind someone's ideas. The markets inevitably do crazy, unpredictable things, which are going to test you psychologically, and the only way to make it through these difficult periods is to have absolute faith in your ideas."

Weinberger, Heimark, Struve, Solo, and the other O'Connor partners have computed the odds on Prediction Company succeeding. Solo is the most pessimistic, at one out of five. Weinberger is the most optimistic, at fifty-fifty. These are not great odds. But everyone agrees, given the potential payoff, the bet is worth making.

David Weinberger should be recorded in the annals of finance as the original rocket scientist on Wall Street. He was a teacher at Yale University and a former researcher at Bell Labs when he decided in 1976 that he preferred gambling as a profession. Weinberger's jump from academia to Wall Street came so early in the game that he had to undergo a battery of psychological tests to prove he wasn't crazy. Back in the 1970s, before Wall Street became a fashionable place to get rich, academia and business were such radically opposed worlds that there had to be something suspicious about a Ph.D. whiz kid giving up an Ivy League teaching and research career for life on the Street.

Weinberger, the son of an engineer, was born in 1947 in Irvington, New Jersey. By eighth grade, without having taken a formal math course, he was tutoring college kids in calculus. He began winning state math tournaments, but he was also good at sports, playing varsity baseball and soccer. He went on to study math and physics at Princeton, where he mastered game theory and wrote a senior thesis on random number generators. It was also at Princeton that Weinberger developed his taste for gambling. He was an avid poker player and bridge player who became an equally avid blackjack card counter after reading Edward Thorp's *Beat the Dealer*.

Before starting graduate school at Cornell University in operations research—a blend of game theory and applied mathematics that was in vogue during the 1960s—Weinberger got married and took his new bride on a honeymoon gambling trip to Puerto Rico. Here in 1969 he played his first hand as a professional card counter. He won.

After four nights of play, his two-hundred-dollar stake was up seventy dollars, which represents a tidy thirty-five-percent return on his investment.

Weinberger was hooked. He became addicted to running the numbers, looking for an edge. He bought computer time on the Cornell University mainframe to analyze card counting schemes. Several of these schemes were field-tested during gambling trips to Las Vegas and Tahoe. Then came betting on football games, college basketball games, NCAA tournaments. Weinberger began assembling a massive database of point spreads. He wrote computer programs to map these spreads into probabilities. He moved into sports betting, "securitizing" tournaments and buying and selling "shares" in teams that were "traded" by Nick the Greek in Las Vegas or his counterparts in London. "I like being smart and finding ways to win," Weinberger admits. "I'm a beat-the-game kind of person."

His original foray into the stock market was again prompted by Edward Thorp. Weinberger scraped up two thousand dollars, the minimum required for opening an account, and walked down to the local broker in Ithaca. Following Thorp and Kassouf's advice in *Beat the Market*, he was looking for warrants, which are a kind of option, to sell short. The broker had no warrants to short, but Weinberger was soon dabbling in options and other statistically oriented approaches to playing the market.

"I'm not a gambler in the sense of liking the action for the sake of the action," he says, drawing a distinction between calculated risk taking and raw gambling. "I like to analyze a game and find a way to beat it. I love uncertain environments because that's where the opportunities are. That's where you can out-analyze other people."

After finishing his doctorate, Weinberger worked at Bell Laboratories for a couple of years. He was designing algorithms for scheduling telephone installers when he got a phone call from Yale University asking him to come teach a computer science course in algorithmic complexity. It was at Yale in 1976 that the twenty-eight-year-old Weinberger, the married father of a two-year-old son, decided he was in the wrong line of work.

"Science would make a nice hobby, but beating the game was what

I wanted for a job. I thought Wall Street might be a good place for me, but I was afraid of it. I believed it was full of shysters and con men who could lift your wallet right out of your pocket. I went to the library to read books about Wall Street, and they all said the same thing: Everyone in the business is a crook."

Weinberger screwed up his nerve and began knocking on doors. Wall Street at the time had none of its 1980s glamour, and it was mired in the long recession that followed the Arab oil crisis. Among the people willing to talk to him, all seemed quite decent, and one seemed both decent and smart. This was Robert Rubin, who later served in the Clinton administration as secretary of the treasury. Rubin at the time was a young partner doing risk arbitrage at Goldman Sachs. Head of the options area, he was willing to hire Weinberger as junior risk arb trader, but only after sending him out for a day of psychological testing.

Weinberger's training period consists of wandering around the trading room for a week. Then he is instructed to start buying and selling junk bonds, which is a new business for Goldman Sachs. Weinberger is supposed to sell a quarter million dollars' worth of "Crane 8s." Crane Company has taken over another company and financed the buyout with junk bonds, in this case, an eight-percent coupon maturing in ten years.

"I was scared shitless," Weinberger remembers of his inaugural trade. "Here I am a scientist by nature, and I have no clue what these things are worth. I've been watching people for a week, asking them, 'How do you decide to sell at that price?' and their answers don't make sense to me. It's nothing I can get my mind around."

There is only one person in the world who knows what Crane 8s are worth. Weinberger gets on the phone and dials Michael Milken, the king of junk bonds. "So how are you making the Crane 8s?" he asks Milken.

This is trader lingo that Weinberger has picked up during his week on the trading room floor. The dealer, not knowing whether Weinberger is a buyer or a seller, is supposed to quote him both sides of the market, with a relatively tight spread. But in this case, Milken *is* the market, and he knows exactly why Weinberger is calling. To get rid of his junk. This is why everyone in risk arbitrage calls Milken.

Junk bonds exist in the netherworld of "fallen angels" or "deep-discount" bonds, whose credit rating is about to slip off the bottom of the charts. But if they manage not to default or, in some cases, even crawl their way up from junk to investment grade, they give an elevated rate of return that more than amply repays their elevated risk. This fact was discovered by Milken in the late 1960s when he was a student at the Wharton School of Business. Trading from an X-shaped desk in an office on Rodeo Drive in Beverly Hills, Milken single-handedly created an entirely new bond market for financing green-mail, hostile takeovers, and other excesses of the "greed-is-good" Reagan era.

Milken sold Weinberger's Crane 8s, like the rest of his "Chinese paper," through a network of investors who believed what he was preaching: that junk bonds are undervalued because people worry too much about default. Unfortunately, the prices of these bonds collapsed during the S&L crisis in the early 1990s. The momentary demise of the junk-bond market cut Milken's customers off at the knees. Many of these people were directors of America's savings and loan associations. The S&L crisis, the worst banking scandal in American history, would drive them out of business and send several of them to jail. The total bill for Milken's experiment, charged to the American taxpayer, will be about nine hundred billion dollars. Milken himself joined his colleagues behind bars in 1990 when he pleaded guilty to insider trading and stock manipulation. He would spend two years in jail and pay a six-hundred-million-dollar fine.

Later, Weinberger screwed up his nerve to make another call, to Ed Thorp, who soon became his best customer. Having made his own conversion from academia to business, Thorp was running his Princeton/Newport hedge fund, trading options and other exotic paper from which one could arbitrage a profit. As practiced by Japanese rice merchants and speculators in Dutch tulip bulbs, arbitrage was the art of buying and selling the same thing in different markets. Arbitrage today is the art of buying and selling relationships among speculative instruments. Instead of rice and tulips, one trades stocks and options on stocks or statistical hunches about how the S&P 500 index should relate to the S&P 100 index. This new kind of arbitrage developed with the invention of the handheld calculator and desktop computer. When Wall Street began attracting the likes of Weinberger and Thorp, arbitrage entered its golden era.

Working his way up from junior arb clerk to head of the options area at Goldman Sachs, Weinberger was present at the creation of the derivatives market. When he talks about Wall Street, he sees in his head a picture of what he calls "market space." In this space is the time value of money and all the other probability assessments that go into deciding the correct price of an investment. The space itself is made up of all the investments available in the world—warrants, options, bonds, stocks, futures—which interact with one another in a dynamic matrix that changes instantly with every tick in the market.

The Chicago Mercantile Exchange began trading currency futures in 1972. The following year, the Chicago Board Options Exchange began trading options, futures, and other financial derivatives. "These are critical for completing the mathematical space in financial markets," says Weinberger. Capitalism, by its very nature, is continually expanding to fill this space. It is an emergent, self-organizing system that will occupy whatever financial ecologies are available to it. After currency futures and stock options, the next frontier to be colonized by capitalism was stock indexes. These allow bets to be placed not on individual stocks, but on groups of stock or the entire market itself.

The first white-collar operator of a stock index game was the Kansas City Board of Trade. In 1982, they began selling futures and options on something called the Value Line Index. At the time, Dow Jones & Company would not let anyone bet legally on the Dow Jones Industrial Average. Nor would Standard & Poor's let its name be used for betting on the S&P 500. So Kansas City had to content itself with a lesser-known index invented by an investment magazine.

At the sound of the opening bell on the first day of trading in the Value Line Index, David Weinberger invented stock index arbitrage, which consists of buying or selling stock index futures in one market and selling or buying the underlying stocks in another. Working the phones and barking orders to his traders on the floor, he bought or sold futures contracts in Kansas City and sold or bought the underlying stock in New York. The Value Line Index, which was meant to reflect the up-and-down movement of the New York Stock Exchange as a whole, was arbitraged against a representative basket of stocks "whose composition I made up by the seat of my pants," he says. Stock index arbitrage, which is now a billion-dollar game, caught on so fast in Kansas City that the Chicago Mercantile Exchange quickly scrambled to open its own S&P 500 futures pit less than a year later.

In the Weinbergerian view of the world, the basic law governing financial markets is that their mathematical space will tend to get increasingly abstract. First come the primary markets. General Motors sells stock to the public, and GM uses the money from this stock offering to build factories and make cars. Then come the secondary markets. These are stock exchanges where shares in GM are swapped among speculators, grandmothers, pension funds, and anyone else who wants to buy or sell them.

What productive value lies in the secondary markets? None. Does the money involved in stock market speculation aid General Motors directly? No. But on the other hand, if General Motors wants to raise money by issuing more stock, the secondary market lets it know the price at which this stock can be sold. Investors are willing to step forward and buy this stock because there exists a hassle-free mechanism for converting it back into dollars. The justification for the parasitical enterprise known as stock market speculation is that it works. It greases the skids that provide the capital for capitalism, which is why stock markets are being installed everywhere from Ho Chi Minh City to Moscow.

In the decade between 1972 and 1982, with the introduction of trading in currency futures, options, stock indexes, and other derivative instruments, the mathematical space of capitalism was further abstracted by the invention of what might be called the tertiary markets. This is not General Motors making a stock offering. It is not people betting that the price of this stock will rise or fall. It is a bet on the betting in the secondary markets, where "General Motors" is no longer a company making cars, but one number in a pot of five hundred numbers whose daily average either rises or falls.

None of the money bet purely on a stock market index goes directly into a stock market, and none of the money bet in a stock market goes directly into a company's coffers, which makes betting in the derivatives markets doubly parasitical. Aristotle, the Gospels, the medieval Schoolmen, and John Maynard Keynes all agree that using money to "breed" money, as Aristotle says, is "the most unnatural way of enriching oneself."

But the increasingly abstract, ever-more-arcane bets in the tertiary markets are doing their own part to grease the skids of capitalism. So heaven help the newly appointed financial czar in Ulan Bator who, not grasping this fact, tries to prevent his citizens from buying derivatives. "Money goes where it is wanted, and stays where it is well treated, and that's all she wrote," growls former Citicorp chairman Walter Wriston; he claims that most of the investment capital in the world is now "stateless money," no longer controlled by the national treasuries and central bankers who used to regulate the flow of money in financial markets.

"The secondary markets are critical to the proper functioning of the primary markets, and the tertiary markets have become critical to the proper functioning of the secondary markets," says Weinberger, who sees these markets as nothing more than contingent probabilities, each of which is useful, in one way or another, for helping investors hedge out interest rate exposure or market fluctuations or other kinds of risk. The key word here is *liquid*. The faster money gets put into play, the more of it there is on the table. "Each secondary market makes the one from which it is derived more liquid," Weinberger explains. "But I can assure you, no one in this office is patting himself on the back, saying, 'Boy, aren't we doing a great service for the economy.' That's not what brought us here."

It is the irony of the markets that greedy individuals, obsessed by the numbers and driven to bet on anything that moves, will somehow coalesce to form the capital flows out of which General Motors builds a factory and puts people to work. Driving this process is the invisible hand of Adam Smith. This fundamental force—which today is called a complex adaptive system—somehow manages to transform private greed into public good. Since the markets are always busily filling up the mathematical space available to them, this fundamental force is already engaged in creating the quaternary markets, which are options on options and doubly derivative bets on global indexes of indexes.

Weinberger allowed himself to be hired away to Chicago in 1983. His replacement in the equity department at Goldman Sachs was Fischer Black, the University of Chicago economist who had invented the Black-Scholes options pricing model a decade earlier. Black's arrival in New York in 1984 is often cited as the date when rocket science first arrived on Wall Street. But Weinberger had already been there. The only difference is that when Black accepted the job at Goldman Sachs, no one submitted him to a day of psychological testing.

Weinberger's first assignment in Chicago is to move O'Connor & Associates into stock index arbitrage. He proves such a steady winner in every game he plays that he emerges in 1986 as comanaging partner of O'Connor, effectively running the whole show. Three years later he changes direction again. He suddenly announces that he is quitting daily trading and taking a leave of absence. He devotes a year to studying physics. "I found a lot of brain rot," says the former prodigy. He takes up Formula One racing. He buys a blue Porsche Carrera and spends a lot of time at his ski lodge in Utah. Still restless, he goes back to work part-time in 1991. He is pacing the trading floor looking for a good game to play, when Craig Heimark tells him about these people he has met from Prediction Company. This is exactly the kind of long shot Weinberger is looking for. After his emergency visit to Santa Fe, he knows he has found the team that is going to help him "out-analyze the competition." He is back in the game.

Not only does Prediction Company remind Weinberger of himself when he was a neophyte in the business world, but they also mirror what O'Connor itself is going through as they negotiate a deal with another business partner. Weinberger and his colleagues are thick in discussions that will lead to O'Connor being bought out before the end of the year by Swiss Bank Corporation. It is the deal of Weinberger's lifetime. A rumored two-hundred-and-fifty-million-dollar sale which will make O'Connor's twenty-three partners very rich. It is also a financial revolution that will lead to the trading operations at one of the world's largest banks being taken apart piece by piece and rebuilt from the ground up with O'Connor technology.

The Merc and Chicago's other exchanges are big players in the world financial markets, but they are not as big as the banks that dial deals among themselves. Already by 1986 over-the-counter trading in the interbank market had begun to replace the shouting taxi drivers down in the Chicago pits, and each year after that the banks' market share crept higher. Interbank trading in derivatives is a high-stakes game that not everyone can play. It requires deep pockets and a high-quality credit rating. Without this pedigree, the world's dozen major banks are not going to answer their phones to do a deal with you.

O'Connor had ridden an explosive growth curve from its early days applying the Black-Scholes pricing model to the options market. The company made money hand over fist in the 1970s because no one knew better than the mathematical whiz kids on LaSalle Street how to value these pieces of paper. As more people caught on, O'Connor got smart in other ways. They started engineering "risk-neutral" investments. These are balanced portfolios of stocks and options, where the risk in one area zeros out the risk in another. Then they moved into risk arbitrage, profiting from the mercurial swings in stock prices that are provoked by buyouts and corporate takeovers.

Another round of explosive growth came with new kinds of options trading. O'Connor led the pack in trading options, not just on stocks, but also on bonds, foreign exchange, and stock indexes. The company grew to six hundred employees working in a computer-intensive environment where statistics ruled. Their training rooms for new hires were called "Black-Scholes" and "Silicon Valley." Wherever the mathematical space of the markets was expanding, O'Connor was prepared to rush in with its pricing models and split-second execution. "A lot of the company's success was just riding the curve of opportunities as the financial world became more complex," says

Weinberger. "There were just that many more places for us to go and be smart."

As O'Connor moved into swaps, floors, collars, swaptions, look-backs, and other exotica in the ever-expanding world of derivatives, it acquired a growing need for capital. Exchange-traded derivatives are guaranteed by clearinghouses attached to the exchanges where the contracts are traded. But over-the-counter instruments are guaranteed by nothing more than a contractual handshake between the participating institutions. Derivatives trading is risky business. Prices on out-of-the-money options can triple in an hour. A single forex swap can involve billions of dollars, and O'Connor was finding people shy about shaking its hand on these big deals.

Weinberger makes the case to his fellow partners. "We have to work jointly with a large financial institution or watch O'Connor shrink and ultimately disappear as our markets dry up. We have plenty of capital for dealing on the exchanges, but not for the interbank markets, where the business seems to be heading. There is no way General Electric and other big players will come to us directly for this kind of business. For all they know, we could close up shop and head to the beach next week. They want to deal with a Morgan or a Bankers Trust. Morgan might then come to us to lay off their risk, but they'll insist on getting their cut of the pie. Only with the credit rating of a bank behind us can we get into the interbank markets where the business is going."

O'Connor tries to build a strategic alliance with First Chicago Bank. O'Connor does wonders for the bank, but the bank is too small to do much for O'Connor. This is when a lot of Japanese bankers are spotted up on O'Connor's balcony. Then one day the Japanese are replaced by tall men in pinstripe suits, and in 1989 O'Connor announces it will begin a five-year trial relationship with Swiss Bank Corporation. SBC at the time was the world's twenty-ninth largest financial institution, with one hundred and fifty billion dollars in assets. In exchange for use of these deep pockets, O'Connor will roll its technology into all of SBC's trading rooms.

Schweizerischer Bankverein, based in the medieval city of Basel, was formed in 1896 through the merger of two other banks in

Rhenish northern Switzerland. Acting less like a corporation and more like a confederation of independent principalities, SBC maintained lots of autonomous branches, each responsible for its own investment decisions. This approach worked well until the 1980s, when an increasingly integrated world economy made it look not only outdated but also dangerous. Realizing that its autarchic structure made it difficult, if not impossible, to figure out the bank's global risk position in real time, SBC, like its two big rivals, Union Bank of Switzerland and Crédit Suisse, launched an effort to remake itself. The financial landscape was shifting under their feet. The profit to be made from lockboxes built under cobblestone streets was nothing compared to successful moves in the derivatives market. The Depression-era barriers keeping American banks out of the financial markets were tumbling down, and either the Swiss learned to play these new games, or they risked becoming a quaint remnant of Switzerland's glory days in the banking business.

The problem at SBC was that they had limited knowledge of how to operate in these new markets. "Their clients wanted them in the options business—to cover their exposure—so SBC had to provide a service about which they knew next to nothing," says an O'Connor official who wishes to remain anonymous.

"Knowledge of derivative trading and risk management has grown out of the United States, not Switzerland or Germany," confessed the head of equities trading at SBC. So foreign was the subject to Rhineland Switzerland that the University of Zurich at the time had no full professor of finance. SBC was a market maker at SOFFEX, the Swiss Options and Financial Futures Exchange, opened in 1986, but this limited local experience needed enhancement if the bank was going to play globally.

The assignment for figuring out how to function in the new financial world was given to a young bank executive with a wry smile named Marcel Ospel. A hometown Basel boy, he had worked briefly for Merrill Lynch in London and New York, and he knew how the Americans did business. "The capital markets of the 1990s will be driven by derivatives and risk management products," Ospel advised his bosses in 1989. He went on to outline a plan for SBC to survive the decade and emerge as one of the world's ten most profitable banks. His plan was deceptively simple, but quite bold for a bank that was still conducting its board meetings in the local Swiss German dialect. SBC, said Ospel, should find the world's most talented derivatives traders and buy them. Target number one was an options powerhouse in Chicago called O'Connor & Associates. There were other targets in New York and London that Ospel would eventually lead the bank into acquiring, but SBC should begin by buying the risk management technology it needed to survive as a major bank.

"We were talking about building a jumbo jet," said Ospel. "Then we decided, let's not build one, let's learn to fly one." Swiss Bank inaugurates a ten-year "strategic alliance" with O'Connor in 1989. The Americans roll their technology into SBC's three hundred and thirty-nine branch offices in forty countries. They set up operations in Tokyo, London, and Zurich and make the bank seventy million dollars in their first year of operation. O'Connor, in the meantime, is happily selling its services around the world as a major player in the options and futures markets. The arrangement is working so well that SBC comes back in year two of the deal to push Weinberger, Struve, and their partners into a buyout. These negotiations are taking place at the same time that O'Connor is talking to Prediction Company. In one conference room Weinberger is playing the little guy who wants to maintain his independence. In the next conference room, he is sitting on the other side of the table as the big guy doing the deal with Prediction Company.

Although O'Connor is being bought out by Swiss Bank, the process, in many ways, is working in reverse. A blue jeans and T-shirt Chicago firm, whose average employee is still in his twenties, is taking over the investment banking division of Switzerland's third largest bank. O'Connor's twenty-one active partners are all slated to become managing directors of Swiss Bank. They will move to London or Switzerland to manage key parts of the operation, or carry these parts back to America. Craig Heimark will run information technology with responsibilities worldwide. Andy Siciliano will handle FX. David Solo will assume responsibility for interest rate trading. Inaugurating the start of a new era in banking history, SBC will be organized to run itself across borders on a global basis.

Within a year of O'Connor's being sold to Swiss Bank, SBC will emerge as one of the top twenty players in the multitrillion-dollar derivatives market. Its foreign exchange options business jumps a thousandfold, and SBC's share of the market goes from one to twenty percent. The bank is named Options House of the Year and Swap House of the Year and ranks number one in a customer poll of institutions trading derivatives. *Risk* magazine names Swiss Bank "preeminent in currency options." SBC wins eight of the magazine's ten categories in the 1993 *Risk* rankings for interest rate swaps, currency rate swaps, options, swaptions, knockout structures, and equity kickers. SBC is beginning to dominate markets around the world, including mainland China, where the bank introduces the first derivatives in 1993. By the end of that year, the *Financial Times* is reporting an eighty-six percent increase in Swiss Bank Corporation's annual net profit.

"Within no time we had our original investment paid back," comments a smiling Marcel Ospel. For his foresight in securing this golden goose, Ospel gets a seat on SBC's executive board and emerges five years later, at age forty-five, as Swiss Bank's youngest chief executive officer.

O'Connor's triumph is not without cost to the casual firm that once prided itself on having a "flat" organizational structure. Where previously all the paperwork had been kept in Struve's head, and no one monitored the free food and other perks that flowed liberally across O'Connor's trading floor, the company is forced by Swiss Bank for the first time in its history to draft a formal budget. Apart from this minor irritant, O'Connor is undergoing another fundamental change in how it does business. Weinberger does not breathe a word of this during his discussions with Prediction Company, but O'Connor is in the process of abolishing its independent research department. Researchers will be moved into regular business units or into a technology department that combines research with business systems.

From the perspective of O'Connor's reorganized scientists, Prediction Company is a stealth operation out in the desert. It is a black-box research group that O'Connor hopes the Swiss will support long enough for it to develop the next insanely great idea for beating the world financial markets. Weinberger chuckles at the thought of what happens next. Prediction Company comes roaring out of the desert, both guns blasting. It cracks the markets and gets away with SBC–O'Connor's next big conquest. Money starts flowing like manna into everyone's pockets. Another game is beaten by out-analyzing the competition.

Cities of Gold

There is no reason to assume the existence of shortcuts.
— JOHN VON NEUMANN

Do I think machines will ever think? You bet. I'm a
machine, and you're a machine, and we both think,
don't we?

— CLAUDE SHANNON

While Jim McGill and David Weinberger are hammering out a contract in Chicago, the crew in Santa Fe is hustling to get their predictive models online. July 1, 1992, is their self-imposed deadline for
going live. After a hundred days' forced march through a swamp of
financial data, they are going to emerge on the other side with a real-
time order to buy or sell. Buy or sell what? They don't know yet, apart
from the fact that it will be futures contracts, which are widely traded
all over the world with plenty of available data. Futures exist on many
different things, including stock indexes, interest rates, currencies,
and commodities. At this point, they all look promising, with plenty of
inefficiencies and hidden patterns ripe for recognition.

Down on Griffin Street, the predictors shout from room to room in
the little house that serves as their global headquarters.

Have we factored leap years into our models?

Where is the Euromarket?

Do they celebrate Christmas in Singapore?

Every day someone discovers a new model that looks better than yesterday's model for predicting the price of crude oil in New York or Deutsche marks in Chicago. Each new model demands more data, and in the furious scramble for numbers, one has to know precisely how many trading days exist in London and whether market x, when it opens, already knows the price in market y. Legion are the ways to trick oneself into thinking that one is predicting the future when, in fact, one is merely peeking into it by accident.

Imagine trying to predict the price of the British pound tomorrow. Hundreds of variables measure the relationship of the pound to itself and other currencies, and each of these relationships can be analyzed. A sophisticated model crunching through hundreds of data streams that collectively produce millions of data points can draw a line through every one of these points. But a model like this is useless for prediction. It has no ability to generalize. The idea is not to mimic the past, but to use it as a tip sheet for predicting the future. Even on a supercomputer, a model fitting lines through hundreds of data streams will be as slow as sludge. Things begin to speed up again when the number of truly significant variables is whacked down to a dozen. A model predicting the price of the British pound tomorrow will want to know the price today. It will want to look at long-term interest rates, but it can live without knowing the price of soybeans.

Down at the Science Hut one of the people responsible for this judicious pruning of data streams is Tom Meyer. Leaning against his computer is a tower of books on global markets and finance. As he riffles through one of these texts, he chuckles to himself, "I came to Santa Fe with software for modeling turbulent fluid flow. The next day I was using it to make market predictions. That's how I became an analyst."

While Tom works on model building, Joe takes on the task of cleaning up Tom's computer code. Computer programming is like handwriting. It reveals one's personality traits. "I don't like people to look at my code because it's so ugly," Tom admits. Some computer programs are elegant and compact. Others are florid, or cobbled together out of great chunks of boilerplate. Tom works with his own private code, which he confesses is "pretty much useless to anyone else."

No one at Prediction Company is a professional programmer. So this is another assignment whose rules they are making up as they go along. They have decided to be egalitarian about the task. Since no one likes to write code, everyone will do it. They assume that computer programming, like finance, telecommunications, nonlinear forecasting, and everything else required to build a global trading system will succumb to physics chutzpah. Physicists are trained to explain all physical phenomena, and they extrapolate from this bedrock knowledge out to the rest of the world by means of physical intuition. This is the combination of theory and street smarts that makes a good physicist good.

"A physicist is a mathematician with a feeling for reality," Norman explains. "We're supposed to understand the basic laws of physics well enough to know what's going on, even in peculiar situations. Doyne is particularly good at making these first-order approximations and sorting out effects, but physical intuition is what training in physics is all about."

Norman offers the following test. Suppose you're holding a helium balloon in an airplane. The airplane takes off. The thrust pushes you back in your seat. What happens to the helium balloon?

It moves forward. Why? Because helium is less dense than air. While the dense air is moving backward, the lighter helium, which can be thought to have negative mass, is moving forward. If you flunked this test, forget physics. And maybe finance isn't the right career for you, either.

While Tom and Joe are tweaking up learning algorithms and Doyne is creating the trade engine, Stephen, John, Tony, and Helen Lyons, who is being groomed as the company's trade accountant, are engaged in building something called the pipeline. It is one thing to design a successful model. It is another to deploy it. This requires data feeds for pumping numbers into the model and a mechanism for getting these numbers spit out the other end as predictions. These predictions then have to be implemented, either electronically or by means of a phone call that sends someone scurrying into a trading pit.

Prediction Company intends to build a fully automated, hands-off system that functions entirely on its own. It will capture market data as they stream into the Science Hut from satellites or dedicated phone lines. Someday it might even include a computerized news-reader that can scan articles in *The Wall Street Journal* for signs of what investors call *market sentiment*. The system will route this information into a mechanism that predigests it and then sends it flowing into a computer model. Out the other end will come a buy or sell order to be executed in Tokyo or London. The pipeline will reconcile accounts, monitor performance, and do whatever else is required of a thinking machine that is dedicated to playing the world financial markets by remote control.

The pipeline will perform another crucial function. Soon after they set up their lawn chairs and folding tables down on Griffin Street, the predictors learn an essential truth about financial forecasting. There is no killer application. There is no single best model for this kind of work. There are good models and, for certain kinds of markets, there are very good models, but the best model is one that keeps generating models and selecting the "fittest" among them. So another feature being designed into the pipeline is a mechanism for nurturing populations of models that are continually being blended and recombined before they are polled for their best opinion.

Doyne is an expert on local linear forecasting. Norman is a whiz at genetic algorithms. But neither of them is an expert on neural nets, which is another powerful tool that everyone thinks they should add to their kit. Tom is assigned the task of becoming the in-house specialist on the subject. Next to his stack of books on finance, he adds another textual tower devoted to artificial neural networks.

Neural networks are adaptive systems designed to recognize patterns and devise general rules for acting on these patterns. They are computer programs comprised of simple, interconnected processing elements that resemble neurons in the human brain. In fact, they bear only a passing resemblance to the human brain. But the metaphor is attractive, and neural nets are effective, so who is complaining? Their adaptive quality makes artificial neural networks—you will notice that the word *artificial* is commonly omitted—good at a variety of tasks. These range from handwriting analysis and speech recognition to robot control and stock market prediction.

If the data being modeled are snowfalls in the Rocky Mountains, the neural network should produce an undulating curve, which indicates that more snow falls in winter than summer. The model does this by a sophisticated form of trial and error. It varies the strengths of connections between individual processors until the input yields the right output. This business of pruning "nodes" (connections), adjusting "weights" (signal strengths), and other arcane tweaks to various network knobs is not accomplished by means of specific instructions. It results, instead, from a general learning program that gets smarter with each pass through the data.

Neural nets offer several advantages over other approaches to machine learning. Since they learn by example, they do not need to be programmed. They are fault tolerant, which means they are good at dealing with noisy or incomplete data. They are gluttons for information, because their interconnected processing units work in parallel, and they are cheap to build. On the other hand, there is no formal method for proving that a neural network will do what you want it to do, and they are prone to discovering patterns where none exist. Neural nets are black boxes. But they perform so well that engineers are now building them into a wide spectrum of machines.

To take another example, imagine trying to predict the price of the Japanese yen as it goes on sale tomorrow in Tokyo. You are relying on a neural net that crunches raw numbers, looking for patterns. What numbers are you going to feed it? Historical data going back to the creation of the yen in 1871 are not particularly useful. Fundamental data on the Japanese economy are published quarterly, which means they are already out of date by the time they appear. Economic data are subject to interpretation. What remains are opening and closing prices, volume, and other technical data.

In the end, your neural net might have six inputs: the yen-dollar rate, the Deutsche mark–dollar rate, the yen–Deutsche mark cross-rate, the yen bond, the three-month interest rate, and a figure describing the Japanese money supply. Open the spigot, and numbers like these will spew from any Bloomberg box or Reuters satellite feed. But pumping raw numbers into a neural network is like pumping raw sewage into a water system. It gets murky fast. Instead, the numbers have to be transformed in ways that enhance their significance. Closing prices, for example, might be combined with trend and volatility indicators. How one transforms the data going into a neural network is the real trade secret that differentiates the work of amateurs from pros. Prediction Company, as it develops its trade craft, will eventually compile a hefty manual of statistical transformations.

The idea for artificial neural networks originated with the publication in 1943 of a paper by Warren McCulloch and Walter Pitts called "A Logical Calculus of the Ideas Immanent in Nervous Activity." McCulloch was a professor of psychiatry at the University of Illinois Medical School in Chicago; Pitts was a teenage genius who would later kill himself. "It is to him that I am principally indebted for all subsequent success," McCulloch wrote of Pitts. They eventually moved to Massachusetts, where they continued their work in neurobiology at the MIT Research Laboratory of Electronics.

McCulloch and Pitts were trying to understand mental activity as a form of information processing. Their paper gave the first hint that a network of very simple logic gates could perform complex computations. This idea was seized by John von Neumann as key to his own research in computers, and it led to a flurry of activity that culminated in the 1960s with the development of the *perceptron*. This was a simple neural network machine built by Cornell University psychologist Frank Rosenblatt. Perceptrons were meant to emulate the parallel processing of the human brain.

Opposed to Rosenblatt was his former classmate at the Bronx High School of Science and arch-enemy Marvin Minsky, who was also working in cognitive science, but under the banner of what he called *artificial intelligence*. Minsky had no interest in imitating the brain. The best way to get results, he said, was to establish an arbitrary set of rules for manipulating signs that refer to concepts and objects. Of these two competing models—perceptrons and AI—one was "natural," with parallel processing designed to mimic the brain; the other was "artificial," with serial processing meant to maximize the search capabilities of digital computers.

In 1969, Minsky and his MIT colleague Seymour Papert published a book called *Perceptrons*. The book was so persuasive in its attack on Rosenblatt that it effectively killed all further research in connectionist cybernetics. It also seems to have killed Frank Rosenblatt. He died on his forty-third birthday in a solitary boating accident that may have been a suicide. One account of these events describes Minsky and Papert as "the evil huntsmen who slew Snow White." Another accuses them of killing their sister science in order "to gain access to Lord DARPA's research funds." This is a reference to the Defense Advanced Research Projects Agency, the governmental body that has spent billions of dollars trying to create an artificial mind.

In the early 1980s, the tables were turned when Minsky's approach to cognitive science reached a dead end, and connectionism sprang back to life. This is when several researchers, working independently, discovered a new technique for training neural networks called *back propagation*. Back propagation works by feeding inputs into a multi-layered neural network and taking the outputs, which are often wrong, and feeding them backward through the network, with various adjustments, so that they eventually produce the right answer. These training sessions can be infuriatingly computer- and labor-intensive. "The trouble with neural networks is that they are too stupid," says an unrepentant Minsky.

Neural networks are also prone to what is called *overfitting*. They see patterns where none exist. In our Rocky Mountain snowfall data, if the four highest snowfalls occurred on Thursdays, then our neural network might predict that tomorrow, being Thursday, will be snowy. Meaningless chance has now been elevated into a prediction. These limitations aside, back propagation was a major breakthrough because it allowed neural networks to find nonlinear patterns. Artificial intelligence could now tackle a wide range of real-world applications, like playing backgammon or predicting stock prices, that were previously closed to it.

Before he begins using them at Prediction Company, Doyne is inclined to think that neural networks are a fad whose benefits are overrated. As a forecasting tool, they suck up ten times more computer power than the local linear method that he and Sid Sidorowich developed at Los Alamos. But Prediction Company is leaving no stone unturned. So Joe takes over the Prophet models, while Tom is redeployed into "quantitative analysis," the fancy name given to neural networkers and other mathematically inclined arbitrageurs on Wall Street.

"Quantitative analysis is the financial equivalent of shooting a rapidly moving enemy tank from one's own tank, while traveling over rough terrain," claims one proponent of the new science. So Tom sets to work, hunkered down at his computer terminal, moving fast over the rough terrain of the world financial markets, while trying to get a bead on the yen–Deutsche mark cross-rate and shoot it down with a direct hit from the Japanese bond.

Apart from his trip to Chicago to meet their prospective partners on LaSalle Street, Norman makes another interesting trip in the spring of 1992. Rafael deNoyo is rumored to have discovered a non-linear forecasting method with returns approaching fifty percent. Norman and Tony Begg set out to investigate these rumors. They fly to Houston and drive from there into the horse-breeding country around Richmond, Texas. The rolling hills are deep green and steamy with humidity. In a big ranch house, out in the middle of grasslands and paddocks, they find deNoyo and his two partners. They are sitting in the living room listening to computers, which are outfitted with neural net graphics packages and speakers, barking out commands to buy Treasury bonds and sell the Nikkei.

Shirtless and stripped to their underwear in the afternoon heat, the men are quaffing large quantities of beer and wine while jockeying trades from Chicago to Singapore. DeNoyo's baby daughter crawls across the floor. His wife sits at a workstation writing code. The rest of the room is a bedlam of cheers and high fives as they watch the telltale signs that signal the approaching collapse of the Japanese bubble economy. The crash will wipe out two-thirds of the value of the Japanese stock market. DeNoyo is happily positioned on the short side of this momentous event. Every five minutes a computer screen flashes red and a synthesized voice resembling that of Darth Vader announces, "SELL FOUR JUNE NIKKEI." Out goes a call to a broker in Chicago, who has no idea he is dealing with a bunch of Texans sitting in their underwear. DeNoyo is trading five markets—Swiss francs, the Nikkei Japanese stock index, the "Footsie" London stock index, the S&P 500 American stock index, and U.S. Treasury bonds. He has trained multiple neural nets for predicting each market, and his models tonight seem to be making money.

"It's a Ferrari look, with a VW engine," Norman concludes on closer inspection. But he is impressed that deNoyo, a high-school dropout, is already rolling down the road, while Prediction Company is still back in the shop looking for a steering wheel.

DeNoyo is a hard-drinking, cigar-smoking, ex–weight lifter who is beginning to bulk out as he hits the back side of thirty. A nonstop talker and endless font of ideas, he likes to conduct his business meetings in strip clubs. He is a radical libertarian who has no use for bank accounts or taxes. He gets his political news from *High Times*. The rest comes from the markets, whose data he consumes in copious quantities. "I'm a feed junkie," he admits, referring to the dedicated telephone lines and satellite dishes he uses to keep numbers pumping into his computers. "Putting the suppliers together with data manipulation, these are my secrets."

DeNoyo lights the barbecue out by the pool as he and his guests settle down for a night of well-lubricated shop talk. He is self-taught in nonlinear dynamics, neural nets, genetic algorithms, and other tools in the chaos kit. He has read Norman and Doyne's scientific papers, which confirm his own intuitive grasp of the subject, and he is excited to be meeting one of the masters in person. He wants to check out a few more hunches. Everywhere he looks he sees opportunities. He figures the financial world is more than big enough for himself and Prediction Company.

For his part, Norman is surprised by deNoyo's apparent sophistication. While Prediction Company is building daily models, which trade near the open or close of the markets, deNoyo has scaled his operation down to five-minute models, which can trade a dozen times an hour.

"Intraday modeling is the way to go," he drawls, in his Cajun accent. "But you have to look sharp. There are two ways the alligators can come up and bite you in the ass. Data are coming at you so fast, it's easy to overfit the models. You get smoke in your eyes and start seeing things that aren't there. The other problem with five-minute models is commissions. You trade too often and the middlemen will eat you alive."

"How do you limit your trading?" Tony asks.

"We hold back and wait for a really big move," deNoyo explains. "Then we get up and ride."

Tonight the moves are so big that every few minutes, all night long, a trader dozing in front of his monitor is startled awake by the sound of a computer-synthesized voice telling him to sell the Nikkei. "Hot damn!" he yells, watching the rising gas gauge on his P&L. "You guys are my lucky rabbit's foot."

"Transaction costs are everything," advises deNoyo, settling back in his chair. "We compute them down to the penny. This gives us an envelope we can't move outside of. Commissions and slippage. People tend to overlook these things, but slippage can kill you."

DeNoyo tops off the wineglasses and uncorks another bottle. They are sitting on the porch, where moths as big as grapefruits are flopping against the screens. "The computer is quoting a two-point spread," deNoyo continues, explaining what he means by slippage. "You can buy at twelve or sell at ten. But when you go to sell, the order gets filled at eight. Somebody just put two dollars in his pocket, and it wasn't you. To avoid slippage, you have to work your broker. You have to get him to turn on a dime."

"So why are you trading five-minute models," Tony asks. "Daily models have fewer commissions and less slippage."

"The daily stuff doesn't work anymore," deNoyo says flatly. He rolls some grilled peppers in a tortilla and takes a bite. "There is a whole crowd of trend followers out there. They use the daily data you and I do—high, low, close, volume—looking for breakouts. Basically, every trader in the market is a trend follower."

"That's another alligator that can bite you in the ass," he warns. "We learned the hard way; you don't trade through number days. When the Labor Department is releasing unemployment stats, you stand clear of the market, unless somebody is leaking the numbers to you in advance."

Everyone agrees, as they dine on the veranda, eating grilled peppers and fajitas, while listening to the moths flapping against the screens and the sound of deNoyo's computers barking out commands, that this can be a fiendishly tricky business. It can also be a lot of fun. The pay isn't bad, either.

Rafael deNoyo was born in 1962 into a New Orleans family that dates back to the days when Louisiana, bankrupt after the explosion of John Law's Mississippi Bubble, had been sold by France to Spain. DeNoyo had a stormy relationship with his father, who once punished him by making him live in a tree house for a month. DeNoyo ran away from home at age fifteen, moved to Los Angeles, and started hanging out at Gold's Gym in Santa Monica. There he pumped iron with Arnold Schwarzenegger and other muscle men who spent their days on Santa Monica Beach.

After passing a high-school proficiency exam, deNoyo enrolled at the University of California at Berkeley, where he studied mathematics and economics. He never graduated from Berkeley, for want of a passing grade in freshman English, which he was too busy to retake. He was already involved in trying to beat the market. He had bought a TRS-80 computer for three thousand dollars and tried to use Fourier analysis for predicting market moves. This technique, named after the eighteenth-century French mathematician Jean-Baptiste Joseph Fourier, allows one to break down complex functions into their constituent components. The process resembles the breaking down of musical chords into their simple pure tones. The analogy is so apt that Fourier's technique for approximating functions is commonly called "harmonic" analysis.

After trying to beat the stock market, deNoyo turns to the international currency markets. Here his chosen method is Baysian analysis. This branch of mathematics can be used to predict both the direction and magnitude of various forces, in this case, financial forces. DeNoyo starts hanging out at a brokerage firm in Tiburon, across the bay from Berkeley. He puts all his money into betting against the U.S. dollar and pegs it right on the nose when the dollar, in March 1985, takes a big swing downward.

"I saw myself as a billionaire currency trader who could kiss college good-bye," he remembers. "I was going to get people with Berkeley degrees to fix my Volkswagen."

In a prospectus he publishes for potential investors, deNoyo describes his financial experience as follows: "Mr. deNoyo spent five years (but never graduated) at UC Berkeley where he wasted taxpayer money and is credited with discovering the link between *ceteris paribus* and catered asparagus. Most of what he was taught at Cal Berkeley, particularly the 'comparative statistics' approach, was pure nonsense. In 1985, Mr. deNoyo left Berkeley to pursue a career in foreign currency speculation."

This career begins with deNoyo bankrupting himself playing the stock index options and futures markets on the Chicago Board of Trade. He has to get a job. He moves to New York and goes to work for an investor named Jay Weinberg. Mr. Weinberg is a wealthy gentleman who drives a Rolls-Royce and lives the high life in Trump Tower, until he is convicted of stealing sixteen million dollars in a Medicaid fraud and sent to prison.

Down on his luck again, deNoyo ships out to Australia, where he goes to work for Elders Finance. Elders is the financial wing of a large conglomerate that owns more beef cattle and sheep than anyone else in Australia. It also owns Foster's Beer and a big portfolio of high-risk, highly leveraged derivatives contracts, which deNoyo helps jockey. Returning to the United States in 1989, he is hired by a New York bank to do foreign exchange trading. Within a year, the bank is convicted of laundering drug money and goes out of business.

In the summer of 1990, deNoyo meets a Texas speculator and Rice University lecturer named Wilbur Edwin Bosarge. Ed Bosarge and deNoyo promise each other the world, and soon they are in business together as Frontier Financial, an investment firm that specializes in using chaos theory to beat the markets. In exchange for serving as Bosarge's financial guru, deNoyo is supposed to get a one-third stake in the forty-million-dollar company. "Problems emerged when I arrived in Houston to discover that Frontier Financial had zero assets," deNoyo recalls. "We were behind on the rent and couldn't pay the phone bill."

"He's a pure con man," says deNoyo admiringly of his former business partner. Before starting Frontier Financial, Bosarge had been involved in other questionable schemes. He was running an oil drilling company when the Securities and Exchange Commission charged him in 1983 with fraud and market manipulation. His company went bankrupt, and his bankers were left holding forty-five million dollars in debt. Bosarge capitalized on this experience by moving into investment banking. He talked a good game, and he was even invited to the Santa Fe Institute to lecture on quantitative analysis.

DeNoyo has a history of getting involved in contentious business deals. These are followed by suits and countersuits and sudden changes of address that result in his temporarily conducting business out of a post box in the British Virgin Islands. Soon after he arrives in Houston, deNoyo suspects he is headed for another offshore retreat. In lieu of the moon, which is not being delivered, he prepares to move his computers out of Bosarge's office and relocate them to his living room. Never one to miss a good party—and Bosarge gives the best parties in Houston—deNoyo waits to make his move until after Bosarge's New Year's bash in January 1991. It is a great party. Bosarge sues. By the time Prediction Company shows up on his front porch, deNoyo is contemplating another Caribbean vacation.

The news from Texas provokes a lively debate at Prediction Company. One side wants to continue building daily models, which are simpler to deploy. The other side wants to scrap them in favor of intraday models, which will perform better in the long run, but are harder to build.

While the debate is still raging, the junior partners embark on designing tick-by-tick models capable of trading every second. They are going to put the market under a microscope that will magnify its every move. But to do this they need data. They need live feeds with ticks. They need a big pipeline of numbers flowing into their computers. Tony phones Telerate, a data supplier in New York. Telerate provides trade-by-trade data to companies, usually via dedicated telephone switches or satellite feeds, but Prediction Company has neither of these. So Tony arranges with Telerate to download their data by modem. This is comparable to placing a long-distance telephone call to New York and staying on the line for a month. The numbers start pouring in, but so, too, does the initial five-thousand-dollar telephone bill.

Jim McGill walks into the Science Hut on July 1, 1992. He has been living on airplanes, flying in and out of Chicago, negotiating the contract with O'Connor. He has been checking in regularly by telephone, but he is surprised to find so many people working on tick models when he thought the company was supposed to be getting its daily models online.

July 1 is D-Day. This is the deadline for going live. The place should be hopping with people pushing product out the door. But, instead, it looks like any other morning on Griffin Street. Researchers are writing equations on the company whiteboard while happily devouring one of Helen's raspberry tortes. There is no completed prediction engine. No functioning algorithms. No models about to fire. Instead, there are a lot of salaried employees beavering away on interesting problems that in a few months might result in killer applications.

McGill summons everyone to the conference table for a meeting. He wants them to resurrect the old daily models and shelve their research on intraday forecasting. "I'm pulling the plug on the Telerate data feed," he announces. "From now on, we're going to work exclusively on daily trading."

McGill phones Pelkey in California and gives him the bad news. Prediction Company will not be meeting its promised deadline for going live. Pelkey flies to New Mexico for a lunchtime meeting with Doyne and Norman. "This is business," he tells them. "When you have partners, you don't embarrass yourself by missing deadlines. The basic rule is: *underpromise, overdeliver.* I don't want this kind of stuff happening again."

The predictors get back to work on daily models. They spend the summer testing algorithms, banging together a data pipeline, building an automated prediction engine, and meeting the performance hurdles that are built into both the prenuptial agreement and final contract with O'Connor, which is still being negotiated. Norman is in charge of reviewing this document. It has gone through so many drafts that, by the end of the summer, it sits on his desk in a pile of paper as thick as the New Mexico telephone directory.

The marriage between Prediction Company and O'Connor & Associates is legally consummated on September 29, 1992. The following day, most of O'Connor is merged into SBC. The new lord of the realm is Swiss Bank Corporation, which, along with buying Chicago's preeminent derivatives dealer, has inherited a five-year relationship with a mysterious entity in the New Mexico desert. As expected, the buttoned-down bankers from Basel are too busy drinking congratulatory toasts and running up the Swiss flag on LaSalle Street to pay much attention to their colleagues in Santa Fe. In this case, one person's subclause is another person's lease on life.

Prediction Company is launched! Swiss Bank sends a big computer with lots of software. Norman and Doyne receive their first paychecks since the company was founded. Doyne celebrates by buying himself a half dozen new Hohner harmonicas. Prediction Company throws a big party with two cases of champagne. Doyne is handed a surprise envelope full of Swiss francs—the company's first proceeds. But there is little time to savor their victory. "All we're thinking about now," says Doyne, with a cockeyed grin, "is putting the pedal to the floor."

First Date

There is one problem with computer programs. Getting
them to work correctly is not easy.

—WHITFIELD DIFFIE

"In-sol-u-bility
The strangest place to be
However you persist
Solutions don't exist."

—CLAUDE SHANNON

Prediction Company goes live, at least in theory, with an imaginary
million-dollar portfolio, in October 1992. Helen picks up the phone
and calls "Scott" in Chicago. "I don't even know his last name, but he's
the most exciting man I've ever talked to," she bubbles. "It was like a
first date." Playing the markets from an adobe house in Santa Fe is an
accountant's dream. Helen sees the world in numbers, and the numbers are starting to come alive.

"I can't tell you how excited I am about finally having a real person
to talk to," she says of her daily phone calls to Chicago. "Scott probably thinks I'm nuts. He's never heard anyone so enthusiastic."

To keep from talking his ear off, Helen rations her questions. Is a
market order filled in three minutes or five hours? If you put up five
percent of the money to buy a hundred-thousand-dollar bond, how
does your liability go on the books? As five thousand dollars, or a hundred thousand dollars? "For the past five months I've been guessing,"
she says. "I read everything I could, but basically I have no idea how
to balance the books."

Helen is a vivacious Californian with flowing red hair. She was a disc jockey for nine years before she opened a computer store in Santa Barbara in 1976. Among her first customers at the Byte Shop was Jim McGill. He was just getting interested in personal computers, while founding Digital Sound, his voice-mail company, but he also played guitar, and soon the two of them were getting together, in Helen's words, "to play the blues, get funky, and hang out."

Disenchanted with California as it grew from tropical paradise to endless exit ramps off one big freeway, Helen moved to Santa Fe in 1981. She baked the cakes at Café Fritz, named after her husband the chef, and ran some cabins down on the Pecos River. She consulted for Computerland in San Francisco and taught business courses at the local community college. Then her old friend Jim McGill called. He was looking for someone to do the books and secure the licenses required for Prediction Company to open its doors.

They hired Jenny Cocq to run the office, and when Jenny got pregnant and moved back to Europe, the company hired a soft-spoken, implacably calm woman named Laura Barela. Soon Laura is being trained to take over the job of phoning the daily predictions into SBC. "That was terrific," says Helen, trying to sound upbeat after Laura's inaugural phone call to Chicago. "But the next time, you might want to speak a little louder. We want them to think we know what we're doing."

Starting with their imaginary million dollars invested in currency futures and T-bonds, Prediction Company racks up win after win in daily trading. They are so hot that Norman even wins the two-hundred-dollar wager that David Weinberger is running on the presidential election. Not only does he predict that Bill Clinton will win, he also forecasts the correct number of electoral votes. The predictors gather for a company meeting in the living room and decide to up the ante. Two days before Thanksgiving, they will start playing with real money. Their bets will be on the books.

"I'm treating our trades just like inventory in a retail store," Helen explains. "Where I used to calculate P&L daily, now I do it every ten minutes." She has also learned how to do real-time accounting of exchange fees, clearinghouse fees, and brokerage fees.

Prediction Company has been shadow trading for weeks. Everything has been working brilliantly. But as soon as they flip the switch to live trading, the system explodes. The day before Thanksgiving is a half day at the exchanges. It is also a "rollover" day, when futures contracts convert from one quarter to the next. Because of a bug in the program, Prediction Company's computer is confused by these events. It thinks the switch in contracts is a sudden spike in prices. To add to the confusion, the computer records a half day's tick data and goes looking for more. It keeps backing up in its program, searching for missing numbers. It develops a stutter, freezes in midcalculation, and finally stops working altogether.

Stephen Eubank swings into full hacker mode. He begins scrolling furiously through lines of code, trying to write new programs that will shunt data around the stumbling blocks and get the system back online. His office feels like a hospital emergency room on red alert. By Monday morning, exhausted from having labored through the weekend, Eubank and friends are ready to declare the patient brain dead.

Stephen phones Jim McGill, who is vacationing in Hawaii. "Keep the system running," McGill instructs him, "even if you have to open up the models and pull out the predictions by hand. Whatever you have to do, we're trading today."

Stephen hacks together a manual system. After a last-minute scramble, the numbers are run, and out come predictions that get phoned to Chicago at 12:50 P.M., ten minutes before the markets close.

"We need to live around trading as the centerpiece of our corporate existence," McGill insists. "System upgrades, maintenance, repairs— everything will now be done under the pressure of trading. Once you've flipped the switch, there's no going back."

Later in the week, during a reprieve in the software wars, the researchers meet at Café Romana to regroup. They think they know how to redesign the system, but they can't do it if they have to spend all their time applying Band-Aids. From their perspective, McGill is leading the company in the wrong direction. He is the company suit, Mr. Heavy, the corporate chill over what would otherwise be a sunny collection of intellectual riffraff. With his talk of *Q4 objectives* and *growing* the company, he seems out of place in Santa Fe. He comes across as uptight and starchy. His interests are too tightly focused on money. "He has poisoned the atmosphere by creating a class structure in a company of eight people," Joe declares.

Their criticism extends to the other members of the company's ruling triumvirate. Norman strikes them as too abstract. He is a dreamer, an idea spinner, a grandiose thinker forever distracted by beautiful theories. Doyne is another thinker surrounded by piles of works in progress. He is an intellectual whirling dervish. Whenever he sees a cloud on the horizon, he can't help but seed it with ideas and hope for rain.

In spite of their different approaches to running Prediction Company, its three directors are surprisingly unified. Doyne and Norman stand behind McGill's decision to keep trading. They learned a hard lesson from their experience in Las Vegas. If you have a winning system, the worst thing you can do is be timid about playing it. Eudaemonic Enterprises failed to make them rich, not because they didn't succeed in beating roulette, but because they were afraid to trust themselves. Instead of playing their edge, they hung back. They kept collecting more data and running more tests, when they should have gone for broke.

Adding to everyone's grumpiness is the fact that this month's predictions are less successful than last month's. Instead of *making* money on paper, they are *losing* it in reality. They are learning that financial markets—driven by mass hysteria, speculation, or chicanery—can veer through billion-dollar curves in the flick of an eye. These are not "corrections," as the pundits say. A *correction* implies that the market has a proper price and that something out of alignment is being put right. This euphemism, like *recession* and *depression*, was introduced to soften the scariness of the original word for these events: *panic*.

The numbers are starting to come alive, but with a petulant will of their own. Two startling events drive home this fact. Prediction Company, to feed its craving for data, has installed a dedicated phone line and tick-by-tick data service provided by a company called Market

Vision. Costing four thousand dollars a month, the service pumps real-time and historical data into the modeling algorithms. It also displays, on a big, full-color monitor, a wide array of charts, graphs, moving averages, stock quotes, and other numerical snapshots from financial markets around the world. The monitor is placed on a table in the living room. The graphics are gorgeous. The pictures mesmerizing. All day long people drift out of their offices to stand in front of the machine, where they watch market data evolve like cellular automata taking on a life of their own.

In place of the standard sawtooth graphs that appear in our morning newspapers, frozen in print, the monitor shows moving images that grow outward and transform themselves with every market tick. From wherever the markets are open—and markets are always open somewhere in the world—come pictures that feel like life-forms evolving in the deep blue soup of graph space. The jagged lines rise. They sag. They oscillate, and sometimes they move so fast that nothing appears on the screen, save for a gap where the ground used to be, before it fell out from under your feet. The numbers are so immediate that one can almost hear in them the sounds of the traders out on the floor, as their shouts and wild gesticulations are piped straight to Santa Fe.

Not long after the Market Vision machine is installed, the predictors are standing around the monitor, watching trading in the British pound, when suddenly, the numbers plummet downward. The fall is huge, with the spiraling graph registering losses in the billions of dollars. It looks as if a black hole has opened under England and is sucking all the money out of people's pockets. The speculative onslaught is so enormous that nothing can stand in its way. Not the Bank of England. Not the G7 industrialized nations. Not the central banks of the entire world.

The predictors are watching the work of Hungarian-born, American speculator George Soros, who is breaking the Bank of England. He is leveraging a billion-dollar hedge fund into a ten-billion-dollar bet that England will be forced to remove the pound from the exchange rate mechanism that links it to the other currencies of Europe. In other words, he is betting that England will be forced to devalue their currency. On Black Wednesday, September 16, 1992, Soros wins his bet and walks away from the market with a profit of about two billion dollars. "It is the coup of a generation," declares pseudonymous author Adam Smith. It is also a stunning example of how the modern economy is driven by speculators.

A month later, the bond markets at the Chicago Board of Trade are jolted by another big bet; this time the speculator moving the markets ends up in jail. Again, a handful of people at the Science Hut are standing in front of the Market Vision machine watching the world markets. The British pound is crashing and burning. The Japanese yen is sinking. The German mark is being kicked to the ground. Only the futures market in U.S. Treasuries is drifting upward, which is what one expects when people are fleeing other currencies. Prediction Company is in the pleasant position of having shorted all three currencies and bought T-bonds. Today they were winning the game.

Then something funny happens. The bottom falls out of the T-bond market. It makes a sudden turn south and keeps dropping. The green line tracking its descent begins to gap. Then something even more unusual happens. The market turns around. It shoots back up to its previous position and levels off for a day's smooth sailing.

This near-miss at stampeding his fellow traders into a market panic is the work of Darrell Zimmerman and his partner, Anthony Catalfo. Shortly after the trading bell rang for the opening of business at 7:20 A.M. on October 22, 1992, Zimmerman and Catalfo pushed their way into neighboring pits, gave each other the high sign, and started flooding the Treasury markets with orders. Zimmerman sold twelve thousand five hundred T-bond futures, with an underlying value of one billion two hundred thousand dollars. Catalfo purchased an equally staggering number of put options on these futures. A put option gives the buyer a right to sell something at a later date. This strategy, called a Texas hedge, reflects a belief that the markets are going to drop. It is illegal. As explained by the *Chicago Tribune*, "A trader carves out an enormous futures position, one that exceeds his net worth by a country mile, and then heads for O'Hare International Airport. If the position turns into a winner at the end of the day, the trader retires rich; it it's a loser, he flees to Brazil."

In this case, Zimmerman is arrested before leaving the building, but not before he has driven the firm that was clearing his trades into bankruptcy. It was, reports the *Tribune*, "a day that stands out as one of the most bizarre and harrowing in the history of the 150-year-old market at the foot of LaSalle Street."

Knowing that computer programs made by physicists are not the best way to implement a global trading system, Prediction Company decides to hire a professional software programmer. John Gibson has left for graduate school at Cornell. There is an empty chair in the Science Hut, and they have deadlines to meet. So out goes a posting on the Internet, under *jobs.misc*. Back comes a flood of two hundred applications from Silicon Valley and the Pacific Northwest. Choosing among the dozen top contenders, Prediction Company hires a crackerjack computer expert named Jim Nusbaum.

With a long reddish brown ponytail and Vandyke beard, Nusbaum looks like a Renaissance courtier transposed from Verona to the Valley. He is a member of Generation X, the posthippie cohort that started tinkering with computers in high school, studied computer science in college, and now runs the machines that run the world—at least that part of it which is jacked into the Information Age. Quiet and intensely focused, Nusbaum's specialty is "mission-critical" programming with zero tolerance for screwing up.

One of Nusbaum's previous jobs involved writing the software for a particle-beam accelerator used by medical researchers to destroy tumors. The software had to be bug-free, because a beam that went off-target could destroy the patient along with the tumor. "This is just what we need," declares McGill. "We want someone who can get our system back on target, so it stops decapitating our programs."

Most recently Nusbaum has been working in Portland, Oregon, for Tektronix, a manufacturer of oscilloscopes. Tektronix has been trying, not very successfully, to branch into high-definition television, digital editing, and other advanced technologies. When the company started firing the people around him, Nusbaum thought it was time to bail out. The job at Prediction Company looks "mission critical"—in fact, it looks desperate.

Nusbaum arrives in Santa Fe just in time to begin sorting out the Thanksgiving rollover crisis. This also happens to be the low point in company morale. Tom and Tony find themselves increasingly at odds with McGill over the direction of the company and its priorities. No one denies that McGill did a brilliant job of negotiating Prediction Company's contract with O'Connor, and he is doing an equally deft job of stroking their new bosses at Swiss Bank. Doyne and Norman are also pleasantly surprised to find themselves agreeing with McGill's judgment calls about the business. They think it important to get their models—any models—online for a six-week trial run. What they can learn from live trading outweighs the few thousand dollars they might lose to glitches in an imperfect system. The only debate is over strategy and timing, and Prediction Company's two principal scientists and its president are of one mind. If you want to play the game, you have to belly up to the table.

"The initial polarization that we thought would give us trouble hasn't happened," Norman remarks one morning as he and Doyne are eating breakfast at the Aztec Street Café. They have come to visit the former Coca-Cola warehouse on the south side of the river. In the old commercial district near the railroad yard, the warehouse has three times more space than the Science Hut and would make a terrific new office for Prediction Company. Norman stirs some sugar in his coffee. "On the hard-nosed business side, we feared it would be McGill, backed by Pelkey, against the rest of us on the hang-loose science side. But it hasn't worked that way. You and I and McGill are a pretty cohesive unit, and it's hard to imagine any insurmountable difficulties if the technology works."

"We've had a great year," Doyne agrees. "Our deal is set up. Salaries are being paid. The research is promising. So why do I feel so bad about going to work in the morning?"

"We have 'sick company syndrome,'" Norman suggests. "And it doesn't seem to be getting better."

The pressure is beginning to take its toll. Doyne is suffering from insomnia. Norman's face has broken out in a rash. They keep revisiting the pros and cons of asking Tom and Tony to look for other jobs. The idea is tough, not only personally, but also philosophically. Power and authority will be introduced into what was once a community of equals. "This means we're buying into the management-employee hierarchy in a big way," Norman warns.

"I know," Doyne agrees. "Its the end of our Eudaemonic ideals." *The Eudaemonic Pie* was a story about growing up. Its sequel is turning into a story about wising up.

Doyne goes skiing with his three children during the Christmas break and begins reading *Doctor Zhivago*. He and Letty have moved into a new house, a rambling Territorial structure in the old Santa Fe neighborhood called "South Capitol." The house sits on a dirt road, fronting an acre of land enclosed by an adobe wall. The yard is filled with trees, including four soaring Chinese elms and the rounded dome of an apricot tree. It also holds a fishpond that later, after Doyne rebuilds it, will be home to three brightly colored koi, who live under a waterfall that feeds a small wetlands of cattails and water lilies.

The house itself is a five-bedroom, single-story adobe with turquoise wood trim around the windows and a tier of brick decorating the roofline. Its other typically New Mexican features include *canales*, gutters that stick out from the roof; Pueblo-style roof beams, called *vigas*; *latillas*, which are small wooden slats fitted between the vigas; and two kiva "beehive" fireplaces.

The house had no heat and was tumbling into the ground when they bought it. It took several months to rebuild, but now they are happily installed at the base of the hill that overlooks the State House and other buildings that distinguish this sleepy town as New Mexico's unlikely capital. The move into town allows Doyne to ride his bike to work, and Letty is about to start a new job in Santa Fe. In January, she begins an appointment as New Mexico's Deputy Attorney General in charge of environmental enforcement.

Doyne is still suffering from insomnia, wondering what to do, when one day his head clears, and he knows the answer. He calls Norman and Jim, who have arrived at the same conclusion. They choose January 6 as the day to break the news. When Tom and Tony agree to leave the company, they will walk out the door with two month's severance pay and their accumulated stock.

Once the deed is done, it proves cathartic. Norman's rash disappears. Joe and Stephen get back to work. Jim Nusbaum throws away Tony's original code and rebuilds the prediction engine. At the next company meeting everyone calmly agrees to pull the plug on the original models. The forty thousand dollars they lost in live trading is written off as a learning experience. Now they will move to build the intraday models that look more promising.

"You did the right thing," Pelkey reassures Doyne and Norman, who are looking grim after their meeting at the Eldorado. "Everyone will feel a lot better once you start making some money."

No Banana

Time doesn't flow at a constant rate. It's coming in spurts
and starts, and big globs of crazy shit happen that change
the world utterly.

—WILLIAM GIBSON

In the long run we are all dead.

—JOHN MAYNARD KEYNES

Norman, Doyne, and Jim McGill fly into Chicago on a raw, blustery
day at the end of January 1993. They are here for a quarterly meeting
with their Swiss Bank partners. "Once every ninety days they think
about us for four hours," says Doyne. "These are busy guys."

The predictors, armed with stacks of research reports, go over the
numbers with Struve and Weinberger. Also attending the meeting, via
a satellite video link from Zurich, is Craig Heimark. Heimark is
now in charge of information technology at SBC's investment banking division. He is one of the busy guys, but he has been intrigued by
Prediction Company ever since he heard Doyne's talk at Esther
Dyson's computer conference, and he avidly follows their progress.

Doyne explains why they pulled the plug on their old models. Then
Norman explains how they expect their new models to work better.
No one cares about apologies for a forty-thousand-dollar loss on
experimental trading, but they care a great deal about what happens
next. *Underpromise and overperform* is the mantra Norman keeps in

mind. He gives no date for when the new system will launch, but he and Doyne are under a lot of pressure to deliver the goods.

The meeting breaks up with the visitors from Santa Fe being turned over to three traders for a tour of the SBC trading floor. They are standing near the FX pod, watching a lot of shouting and finger wagging, as someone explains how the *bid-ask* spread is reported in futures trading. This is the difference between the price one pays to *buy* or *sell* a contract. The two are clearly distinguished in stock market quotes, but not in the futures markets. The data feeds running into Prediction Company show ticks oscillating up and down, like crenelations on a medieval fortress, but there is no indication of whether a tick is a *buy* or a *sell*.

Imagine visiting your local used-car dealer to sell him your old Ford. He kicks the tires, points to a dent in the fender, and offers you a hundred bucks. Suppose the following day you are tempted to go back and buy your Ford off the lot. The dealer will point to the low mileage and tell you he can't let the car go for less than two hundred dollars. This is the difference between the *bid*, buying price, and the *ask*, selling price.

To explain how this works in the pits, the three traders act out a little skit. One of them pretends to be a dealer-market maker with two phones held to his head. A broker in the pit is "talking" to him with hand signals, and the other trader, chiming in from the side, is pretending to be customers, like Prediction Company, who are phoning in their orders. The broker is flashing numbers off his chin. The dealer-market maker is yelling into one phone, *Buy at four; sell at six!* and shouting into the other phone, *Will you do it? Will you do it?*

The difference between a four *bid* and a six *ask* is what traders call a two-tick *spread*. "You can expect to get hit for two ticks on every transaction," they explain. "But when the market is moving, the spread can be as high as ten ticks."

Data vendors do not report bid and ask prices for futures contracts. They quote one price, which is the latest trade, regardless of whether it was a bid or an ask. As they watch the skit being acted in front of them, Norman and Doyne exchange a knowing glance. They always knew transaction costs would be significant; now they realize exactly how significant. Futures contracts for German marks cost one hundred and twenty-five thousand dollars apiece. A *tick* is one one-thousandth of a contract, or twelve dollars and fifty cents. Two ticks is twenty-five dollars. Ten ticks is one hundred and twenty-five dollars. For the first time, they realize that this is the threshold over which their models have to climb before they can start making money.

"We encountered a serious blow," Doyne announces on returning to Santa Fe. Everyone is gathered in the living room of the Science Hut for a report on the Chicago meeting. There is barely room for them to squeeze around the table in a space filled to overflowing with Sun workstations, squawking modems, and a laser printer balanced on top of filing cabinets. "Transaction costs are going to be steeper than we thought."

"We should have focused on this earlier," Norman admits. "We weren't intimate enough with the machinery to realize we were going to face this kind of resistance on the floor."

Swiss Bank is trading their book at the lowest rates and commissions available, but Prediction Company has no way to avoid the friction that the market builds into its prices. "This is just the level of coarseness you have to expect in executing contracts," Norman warns.

He mentions a possible solution to their problem, which he and Doyne had developed on the flight back to Santa Fe. "We'll create multiple active predictions and take the average of all of them before firing the model." This approach is time-consuming, and it chews up enormous amounts of computer power, but the bar to success has been raised, and this looks like the best way to jump over it.

At the end of February, Prediction Company moves into the former Coca-Cola warehouse on Aztec Street, a two-story, dun-colored building in an old neighborhood graced with adobe churches and casitas, which shades farther south into trailer homes and weedy lots fenced for horses and chickens. The cultural epicenter of the neighborhood, which hosts twenty restaurants in a two-block area, is the Aztec Street Café. This unassuming coffee bar is filled with dogs and travelers, Dead Head musicians, St. John's students, girls with nose rings, dreadlocked hippies, New Age seekers, visiting Eurotrash, and

SFI researchers, all of whom agree that this is the best place in town for caffeine.

Directly across from the cafe is 320 Aztec Street, which houses a used-clothing store and junk dealer on the ground floor and the offices of Prediction Company on the second. Their new digs are surprisingly spiffy. The previous occupant was a successful photographer, who managed to build in Santa Fe what looks like a New York City loft. Around a big room with sand-washed walls and skylights are a kitchen and bathroom, complete with Jacuzzi, and several smaller rooms with picture windows looking out to the mountains. The atelier includes a staircase leading up to a rooftop gazebo, which provides a sweeping view all the way from Santa Fe down to Albuquerque. To the north rise the jagged red peaks of the Sangre de Cristos and the blasted caldera that looms over Los Alamos. To the east lie the two sacred peaks, Moon Mountain and Sun Mountain, which grace Santa Fe like a spiritual Wonderbra. To the west stretches the smoky-blue mesa that has been carved into a paleolithic layer cake by the Rio Grande. To the south is the Santa Fe Indian School and old mestizo neighborhood out of which rises the heady smell of mesquite mixed with garlic and chili.

Before Prediction Company moves into their new office, Swiss Bank sends out a security expert to examine the building. The predictors are using a dedicated telephone line logged into SBC's computers, and this arrangement has to be secured against intruders. The inspector notes the big windows in the building, the skylights and balcony off the kitchen, the stairs leading up to the rooftop gazebo—all potential security risks. But McGill, with his no-nonsense demeanor and tough-guy technical talk, calms him down with assurances that Prediction Company will cover the windows and skylights with motion detectors and triple lock their sensitive gear in the old, windowless darkroom.

Into this darkroom, which is newly outfitted with an air-conditioning system, independent power supply, and code box, go an impressive array of servers, modems, and routers. Another piece of gear required for telecommuting to Wall Street is a dedicated phone line or satellite uplink. Microwaves bounce from orbiting satellites in a quarter of a second. Photons cross the continent in thirty milliseconds. For the telecommuters in Santa Fe, the markets are nothing more than 1s and 0s scrolled across the cosmos, and the numbers are the same whether you are standing in Timbuktu or the T-bond pit.

A change of venue is proposed for their quarterly meeting with Swiss Bank in August 1993. Instead of gathering on LaSalle Street, seven SBC officials will fly into Santa Fe. This is Prediction Company's first corporate powwow. It is also the meeting at which the predictors are slated to reveal their new models.

Feeling hard-pressed to find a winning system before the Swiss Bankers arrive, Doyne and Norman decide to throw more people at the problem. They need another researcher and two or three more software specialists. New people cost money, and money at this stage of the game is expensive. One has to give away a piece of the company to get it. But failure is even more expensive. So Pelkey and McGill are asked to recruit some technologically savvy investors and tap them for a few hundred thousand dollars. In short order, they have signed up two of the savviest in the business: Esther Dyson, the computer guru, and Robert Maxfield, cofounder of ROLM, a Silicon Valley telecommunications company that was bought by IBM in 1984 for one and a half billion dollars. Zeus Pelkey also chips in another round of financing. This infusion of capital, which comes in exchange for a twenty percent ownership stake in the company, allows Doyne and Norman to go looking for talent.

In April Doyne is attending a conference on neurocomputation at the Snowbird ski resort in Utah, where he meets a fellow conference-goer named William Finnoff. A dark-haired, wiry chap, who always looks ready to hit the slopes or paddle his way through white water, Finnoff is an Olympic-level athlete. He grew up in Steamboat, Colorado, where his father, after making a lot of money running a successful machine tool business, opened a sporting goods store. The store's primary function was to keep the Finnoff family supplied with kayaks, skis, luge sleds, and other sports equipment.

William skipped college to go to work for several years as a derrick hand, drilling for oil both in Wyoming and Alaska's Prudhoe Bay. To prepare for the American luge racing team, which would compete in the 1980 Winter Olympics at Lake Placid, he was racing in the World Championships in Germany, when he thought he might like to go back to school. He moved to Munich, where he finished a master's degree in mathematics and a Ph.D. in mathematics and economics at the University of Munich and then went to work for Siemens as a senior researcher in the neural networks group. One of Europe's largest corporations, with four hundred thousand employees, Siemens is the German equivalent of General Electric. Like GE, it has learned that manufacturing toasters and refrigerators is less profitable than speculating in the markets.

At the Snowbird conference, William and Doyne spend a lot of time together on the slopes. William skis with Germanic precision. Doyne, in Finnoff's words, "skis like forty miles of bad road." But what he lacks in style, he makes up for in nerve, careening with wild abandon down all the black diamond slopes. At the end of the conference, Doyne invites Finnoff to Santa Fe. In short order he has accepted a job offer as Prediction Company's newest research scientist. After giving notice in Munich and moving back to the States, William is scheduled to start work the day the Swiss Bankers arrive in Santa Fe.

Another new hire at Prediction Company is Doug Hahn, a rangy, affable young man who was a friend of Jim Nusbaum's when they worked together at Tektronix. An expert programmer and Mr. Fixit, Hahn's first assignment is to rewrite the software that was thrown together by physicists moonlighting as programmers. "Doug is like the quick massage woman down at Wild Oats," says Doyne, referring to the "wellness" station which is set up in the middle of the local grocery store. "He knows how to put his thumbs and fingers on just the right pressure points." All day long, Hahn pops in and out of people's offices, providing quick fixes for their software. He keeps taped next to his own computer a picture of St. George slaying the dragon.

They moved to Aztec Street in February 1993, but six months later the office is still a hodgepodge of packing boxes. The corners are littered with printouts and bicycle helmets. The furnishings consist of several three-legged chairs and a lot of Sun workstations surrounded by books, offprints, soda cans, and candy wrappers. With everyone else working furiously on model building and programming, Jim McGill is responsible for making the office look presentable. "I know you guys are pretty smart," says the carpenter who has been hired to install bookshelves. "But what is forty-five plus forty-five?"

"Ninety," says McGill.

"That's right," says the carpenter. "So how did you figure that the wall space for the desk in your office was eighty inches? I'm going to have to put it in my vice and shrink it, and in the meantime I'm sending back your desk and buying one that's shorter."

Trying not to laugh, Laura Barela is standing on a stepladder hanging pictures. The studio, which doubled as an art gallery, is laced with track lights that shine down on ocher walls. Everything that goes on these walls looks great. The Prediction Company art collection, aside from one fluorescent Einstein, includes a lot of posters from Doyne and Norman's former lives as physicists. Many of these include surreal drawings by Chris Shaw, artistic brother of Rob Shaw, the eminence grise of the Chaos Cabal. The poster for the first artificial life conference shows a video monitor evolving into a dancing android. The poster for a NATO conference on information dynamics is illustrated with a typewriter, which has grown hands and started to hunt and peck on its own keyboard. Another poster, for a conference on evolution, games, and learning, shows two ectomorphs playing a game of computerized chess.

Also hanging on the wall is a framed cartoon from the *Harvard Business Review*. It shows two businessmen in suits peering into a crystal ball. Along with his suit and tie, one of the businessmen is wearing the plumed cap of a Renaissance courtier. The caption for the cartoon reads, "The Prediction Company will make investment decisions with the help of a computer simulation." This generic remark is similar to saying that Julia Child cooks with the help of a stove. The cartoon reveals how word of Prediction Company's existence is spreading far afield, and how ignorant people are about the nature of their financial forecasting.

The carpenter busies himself installing bookshelves and filing cabinets that define a conference area in the middle of the big central room. He assembles new chairs and fixes the old wooden ones by screwing their legs back on. The mock mission table from Griffin Street is positioned in front of the whiteboard. A magazine rack is installed with *Futures* magazine shelved next to *Global Finance*. Thanks to Laura, wandering jews appear on top of the filing cabinets. Ferns are hung from the exposed metal struts in the ceiling. Indian pots decorate the bookshelves, which hold, in alphabetical order, a reference collection that includes Fink and Feduniak's *Futures Trading: Concepts and Strategies*; MacDonald and Taylor's *Exchange Rates and Open Economy Macroeconomics*; and Trippi and Turban's *Neural Networks in Finance and Investing*. When the last boxes are carted off, and when the kitchen is stocked with Famous Amos chocolate chip cookies and a case of Blue Sky soda, Prediction Company looks like a very sharp outfit.

The night before Swiss Bank comes to town, Doyne and Letty host a dinner party at their house on Madrid Road. It is a warm evening in August, with the sky lit fireball red by the setting sun. Later, the stars pop up, as an inky black night settles over the mountains. The Gypsy Kings are burning up the stereo, but out in the yard the music is mixed with the soothing sound of water trickling into Doyne's fishpond.

Doyne is in the kitchen mixing margaritas. He is talking to Chris Langton about SWARM, which is Langton's new beehive-inspired software for creating artificial life. Langton has brought along his guitar. He and Doyne are two of the four members in a band called the Blue Jayzz, and he is hoping to get in some licks later in the evening.

Letty, wearing a lawyerly blue skirt and white blouse—not her favorite attire—comes home from the attorney general's office. "Every day is a new range war," she says with a shrug, referring to the Intel computer chip factory in Albuquerque and the cattle ranchers and Indians who are fighting each other for the state's limited water supply.

She kisses Doyne and asks, "Have you found a model yet?"

Swiss Bank is expecting them to flip the switch on a winning system, but every system they develop, no matter how well it performs on historical data, falls apart when it moves into the present. The rules keep changing, as if the markets reinvent themselves and start playing a new game every day.

Jim Pelkey drives his black Cadillac up to the front porch. One of Doyne's former postdocs from Los Alamos, who is arriving at the same time, offers to help, but Pelkey insists on opening his car door by himself and unfolding his own wheelchair. He uses his arms to swing his limp legs out from behind the steering wheel, which is rigged to put the brakes and gas pedal within reach of his hands. After tossing a cell phone into a backpack, he is ready for business.

Pelkey rolls himself into the kitchen. It is a big skylit room, which was created by knocking down walls and retrofitting the rear of the house with French doors that open directly into the yard. He begins talking about a recent trip to Hong Kong, where he visited acupuncturists and drank a lot of peculiar liquids concocted by Chinese traditional healers. He is in constant pain from the bullet still lodged in his spine, and he scours the world looking for relief.

The man is superinformed, like a salt solution waiting for the introduction of the slightest thread in order to precipitate a crystal. He can speak with authority about engineering, business management, telecommunications, venture capital, and pain—his new specialty. He is writing a book on the history of computer communications. He is chairman of the board at the Santa Fe Institute. He is a Type A personality, hustling from deal to deal and keeping score, like everyone else in Silicon Valley, by counting money.

Norman, looking dapper after a month's vacation in Italy, arrives to a flurry of *ciao*s. Even while reclining on the shores of Lago Maggiore, he was not immune to tensions back home. He managed to send a stream of models back to Chicago to run on SBC's computers. On returning to Santa Fe, he found his desk heaped with test results from the thousand most-promising models. Now he is sifting through the Italy batches, working until five in the morning, looking for likely winners. After an initial rush of excitement, he realizes that some of the models are too good to believe. He cleans up the data and retests

204 • THE PREDICTORS

them. A few of the models still look good, but none of them, so far, has managed to carry its winning ways from the present into the future.

"Any luck?" asks Doyne.

"Not yet. But it feels great to be sitting here sipping a glass of wine while my computer is off crunching data."

"It's probably crashed," Doyne jokes. "It's telling you you have a 'segmentation violation.' "

"Ah, you know too well," Norman laughs. Then he steps behind the stove and busies himself cooking pasta with a bacon and tomato sauce, while Doyne marinates the meats and fish that are headed outside for the grill.

When dealing with mountains of data recorded over the last five years from markets around the world, their models soar up graphic peaks and zoom down valleys without missing a single bump in the mathematical terrain. But something invariably goes wrong when the models are trained up to the present and set loose on the future. They go off track. They zag instead of zig. What small advantages they offer tend to get wiped out by transaction costs and other charges.

The bogeyman spooking the system is called *nonstationarity*. It's an old nemesis, first confronted back in their roulette days. The term describes how markets, which used to be in one place, can suddenly veer off in a different direction. Statistical behavior changes in such a way that financial data from the last five years might have no bearing on the next five minutes. The markets are a moving target. They kept remaking themselves, switching gears, altering their premises. They are proving to be an elusive quarry, as fickle as the people who run them.

The numbers flooding into Prediction Company look like digits following one another in an uninterrupted cascade of closing prices, opening prices, buys, sells, bids, asks. But what if this river of numbers holds unseen eddies, whirlpools, sudden sheers to the left or right that hide what the predictors have taken to calling *regime shifts*? These are sudden changes in direction where markets—which have been trending upward or downward or otherwise behaving themselves with discrete moves—suddenly veer in the opposite direction.

Behind these regime shifts lie nonlinear dynamics and crowd psychology. Global markets are an opinion poll conducted in real time around the world every second. Shake these opinions, surprise them with rumors or a new idea, and they can turn hysteric or euphoric in a flash.

The collapse of the Japanese Bubble economy and other Asian markets was a regime shift. So, too, were George Soros's successful attack on the British pound, Richard Nixon's decision to abandon the gold standard, the invention of options and derivatives, the demise of the Soviet empire, and whatever else in world affairs moves markets. The big changes look explicable, but how about the little moves or the moves that happen for no apparent reason? Are the markets continually fluctuating, or do they periodically rest on certain attractors? "It feels as if there is a knob turning in the background," Doyne complains. "To the extent that we're always trying to hit a moving target, then statistical averages aren't relevant, because they don't really exist."

Many natural phenomena are nonstationary in their behavior—a fact first observed by British hydrologist Harold Edwin Hurst, who was an employee of the Egyptian Ministry of Public Works between the First and Second World Wars. While studying a thousand years of data from the Cairo nilometer, Hurst discovered that there is no such thing as an average flow of water in the Nile River or an average length of time between major floods. "There has been a great deal of investigation as to trying to forecast the flood; nothing of any practical use has come out of it, so you don't even know one year what the next year's flood will be like," he remarked.

Hurst began looking at other time series of long duration. The oldest natural series known to him was a two-thousand-year record of annual clay deposits on the bottom of Lake Saki in the Crimea. Other useful records come from annual rainfall and temperature measurements, the growth of tree rings, and astronomical data such as sunspots. At the end of his analysis Hurst discovered that many natural events are not random. They appear to have a memory of the past that results in the past repeating itself. Record-breaking floods or droughts tend to cluster together. "There are long stretches when the floods are generally high and others when they are generally low," Hurst observed. The same is true of the markets, where a big move one day is likely to be followed by another big move the next day. Volatility runs in trends.

A histogram of random events will always conform to the same shape. This is a normal Gaussian distribution or bell-shaped curve, with a hump on top that falls away to two tails on either side. These tails represent rare events, such as rolling a hundred cat's eyes in a row or meeting an eight-foot man. But Hurst's graphs were not Gaussian in their distribution. They possessed fat tails, those clumps of data clustered toward the edges of the graph, where the extreme events like droughts and floods occur. Hurst devised a way to measure the deviation of these fat-tailed graphs from their normal cousins. His discovery is a number, now called the Hurst exponent, which measures the probability that one abnormal event will be followed by a similar event. A large Hurst exponent indicates a process that tends to go in runs. Hurst's mathematical term is much appreciated by people who are trying to measure chaos and other nonlinear phenomena. It is also appreciated by the financial analysts who have noticed that market data produce fat-tailed graphs.

Consider the following example. Something is measured for ten years. Each year's measurement is one, which means the average value of these measurements over a ten-year period is also one. Then comes a big event that measures ten. Immediately the average jumps to two. Then in a hundred years there is a freak event that measures two hundred. The average now jumps to four. One begins to see how, for Nile floods and stock market prices, there may be no such thing as a meaningful statistical average. No one knows why market data produce fat-tailed distributions. Is it a sign of chaos? Is it nonstationarity? Is it fractals? Many theories have been advanced, but none proved. This leaves us knowing little more than what J. P. Morgan observed, "Stocks will fluctuate."

"Nonstationarity is not a single animal, but rather a zoo full of many different kinds of animals," Doyne writes in one of his many research papers, marked "Company Confidential." The best way around the problem, he concludes, is to build "ensembles of models" flexible enough to track nonstationary targets. Prediction Company's foreign exchange models, for example, are really combinations of thirty-two submodels. They work on the principle of "overlapping predictions," which are "voted" together before arriving at a decision to buy or sell.

"One of the fundamental truths about markets is that the dynamics are nonstationary," Norman explains. "We see no evidence for the existence of an attractor with stable statistical properties. This is what characterizes chaos—having an attractor with stable statistical properties—so what we are seeing is not chaos. It is something else. Call it an 'even-stranger-than-strange attractor,' which may not really be an attractor at all.

"The market might enter an epoch where some structure coalesces and sits there in a statistically stationary pattern, but then invariably it disappears. You have clouds of structure that coalesce and evaporate, coalesce and evaporate. Prediction Company's job is to find those pieces of structure that have the strongest signal and persist the longest. We want to know when the structure is beginning to emerge or dissolve because, once it begins to dissolve, we want to stop betting on it.

"All we have right now is a mushy picture of what this coalescing structure might be. The picture is neither well formed nor well studied in the realm of statistics or modeling or dynamical systems. It doesn't have a name yet, other than stranger-than-strange attractor. We could call it 'shifting or evaporating chaos,' but this is a contradiction in terms because one of the fundamental truths about chaos is that it has what are called invariant measures. These are the stable statistical properties that characterize a chaotic attractor. So you don't really want to associate our stranger-than-strange systems with well-understood mathematical properties like chaos. Nobody has an any theorems or models for these systems. We're just out there beyond the frontier of what's well understood mathematically."

Chastened, but not vanquished, Prediction Company still has a variety of statistical tricks up their sleeves. The general rule in science is to try the simple things first. One might assume the markets are consistent and treat one day's data like another's. But the markets are not consistent. At any moment they can veer off and start doing something completely different from what they were doing in the past. The pundits talk about *bear markets* turning into *bull markets*, but this shifting dynamic happens at much finer scales and far more frequently than people have noticed. Prediction Company's solution to this problem is to shuffle its data sets like decks of cards. Adaptive modeling is another solution. This involves swapping models as soon as markets begin to change.

Prediction Company is committed to flipping the switch on a live system by the end of the month. No one knows what this system will be, but "Swiss Bank has a gun to our head," Doyne explains to Chris Langton. "They want to start trading now. They think we'll learn as we go."

Norman is heading out the door for a late-night session at the office. He is going to review his thousand models, run the numbers on the big machines in Chicago, shuffle the data sets, tweak the variables, and try one more time to think his way to a solution. "Thanks for dinner," he says.

Doyne pulls out his harmonicas. He has twelve diatonic blues harps for hitting all the keys in "Jail House Blues" and "Treat Me Like a Fool." Whenever Doyne is feeling flush or wants to celebrate something—like quitting his job at Los Alamos—he goes out and buys a new Hohner *meisterclasse*, which now cost eighty bucks apiece.

The Blue Jayzz, with Doyne on blues harp and vocals, Chris Langton on lead guitar, Stephen Pope on piano, and David Johnston, a St. John's student, on rhythm guitar, try to get together and jam every Wednesday night. On a typical evening, Chris Langton, wearing his standard blue jean jacket and denim pants, will wander into the house and sit down to eat a lamb chop. Then Dave will be eating lamb chops and potatoes, and Stephen will be tucking into dessert. After dinner, they plug in their amps and get down to playing the old songs about trains and loneliness and foxy ladies who are going to love them before the night is over. They mix blues and jazz and old Elvis tunes, everything from Woody Guthrie to Taj Mahal, so long as it gets the house rocking.

Tonight the band is down to two members and some pick-up musicians on guitar and piano, but they do a credible job of playing "Moondance" and "You Really Got a Hold on Me." The late-night session ends with Doyne singing a spirited rendition of Cole Porter's "Don't Fence Me In."

> *I want to ride to the ridge where the West commences*
> *Gaze at the moon till I lose my senses,*
> *Can't look at hobbles and I can't stand fences,*
> *Don't fence me in.*

The following morning, when Doyne shows up on Aztec Street, he finds on his desk a graph charting the returns on one of Norman's Deutsche mark currency models. The model has been a lead contender for powering part of Prediction Company's trade engine. The graph for the last five years shows a trend line rising gloriously upward, except for the most recent six months of data, when the line falls precipitously. It makes a halfhearted attempt to revive itself and then lies there writhing from loss to loss. Norman, on finishing his test run at five o'clock in the morning, had scrawled a message across the top of the graph, "NO BANANA—FINAL."

Phynance

Prediction Company's chance of success is not zero, but close to it.
— EUGENE FAMA

Son of man, can these bones live?
— EZEKIEL 37:3

At nine o'clock on a Tuesday morning in August, seven Swiss Bankers arrive at the offices of Prediction Company, where everyone is logged into their computers, casually typing away, as if it is no big deal having two hundred billion dollars in assets walk in the door.

When everyone is seated around the conference table they begin what looks like a board meeting crossed with an academic seminar. The Prediction Company researchers, including William Finnoff, for whom today is his first official day at work, project onto the wall models full of equations, Sharpe ratios, drawdowns, and other arcana from the new science of phynance.

The Swiss Bank team includes David Weinberger, who has been known to quaff a half dozen espressos at one sitting and not register any effect. Next to him sits stone-faced Clay Struve. He is running the numbers in his head and missing nothing, even while his eyes wander out the window to admire a lapis lazuli sky pearled with nacreous clouds. Next to him is Craig Heimark, the IT maven who first spotted Prediction Company at Esther Dyson's computer conference; since then, he has moved to Zurich and started sporting fashionable European spectacles. The rest of the Swiss Bank team consists of a prosperous-looking chap who directs equities trading; a company troubleshooter who bears a strong resemblance to Bart Simpson; a recent graduate from MIT, who is waiting to become of legal age before he can start trading in the Chicago pits; and a gray-haired gentleman, who used to work for O'Connor's forty-person research department.

On the Prediction side of the table are Doyne and Norman, who look no less rumpled than usual after their late-night labors. McGill is ducked behind his PowerBook typing notes. Stephen Eubank and Joe Breedon look subdued. They are spooked by Norman's "No Banana" graph and not pleased to be reporting bad news. Finnoff, on the other hand, looks quite confident. He is sporting a new goatee. His shirttails are untucked and his sleeves rolled up for action. Sitting back from the table are Jim Nusbaum and Doug Hahn, the two people who will be charged with getting today's bright ideas online. The only key person missing is Zeus Pelkey. The Aztec Street office has no elevator, and he refuses to be carried upstairs. A board meeting scheduled later in the week will be held in a conference room at the Eldorado Hotel, where Pelkey can wheel himself.

None of these people look like sappers plotting an assault on the world financial markets. There are no Masters of the Universe. No Big Swinging Dicks. No one is shouting about "blood in the water" or "ripping the face off" some unsuspecting client. These are not Wall Street cowboys who play the markets with their fingertips, or Michael Milkens who trade on inside information, or voodoo chartists or gamblers. Everyone here has a healthy respect for the laws of probability. They believe that superior ideas and technology are the best way to make money in the markets. They may not understand the fine points involved in Prediction Company's equations, but they know a statistical edge when they see one.

The Swiss Bankers are wearing dress-down-Friday casual clothes, but they manage to look corporate even when they are looking casual.

Every member of the Swiss Bank team, except for the newly arrived MIT grad, is wearing sports shirts and chinos, and boat shoes without socks. On the Prediction side, Norman is wearing boat shoes *with* socks. Doyne is wearing sandals, McGill is wearing sneakers, Finnoff is wearing a pair of German black felt slipper shoes. The predictors have no alligators or polo ponies on their T-shirts, and the rest of their attire ranges from blue jeans to shorts.

The meeting is being held in Prediction Company's "boardroom," a pleasant, skylit space defined by chest-high shelves. Beyond the shelves is the trading station—a handful of computers and modems transmitting predictions out to the markets. All around the room are large windows facing north to the mountains and south to the Rio Grande. The mock mission table has been glued back together, and the wooden chairs are holding firm on four legs, as Doyne steps to the whiteboard.

"We look at the data and let the data speak," he begins, introducing the general discussion that will run until lunchtime. He gives a quick précis of everything involved in building the Prediction Company pipeline. "The pipeline is where you make the learning process and the operating process shake hands," he explains.

Next, Stephen talks about how to find significant patterns in noisy data. "Let's say you want to predict Mississippi flood levels or tomorrow's close on corn futures," he says, projecting a graph that shows data from the great Mississippi flood of 1993 correlated with grain prices in Chicago. "Why is it hard to make these predictions? Because there is a lot of noise in the data and not much information."

He outlines the range of tricks Prediction Company employs for finding structure, or significant patterns, in data. Their method resembles an all-in-one web crawler, which employs dozens of search engines sent out simultaneously to attack a problem. "We're really looking for speed in this," he explains. "Three inputs alone can give you millions of data points to look at."

Joe takes the floor to talk about the next step in the process. Once the numbers have been "cleaned" by Stephen, they are fed into Prophet or other genetic algorithms that sort the populations, mutate the best members, and create a new genome by iterating the data through thousands of generations. "We don't have a lack of models, and there is no problem finding things to put in them," Joe declares. "The problem lies in knowing what to throw out."

William, who is wearing baggy black trousers and a collarless shirt, which make him look like Richard Burton in mufti, takes the floor for a minilecture on neural nets. This is another technique Prediction Company is using to find structure in its voluminous data. Siemens does financial forecasting with neural nets. William helped design the program, and the company, for a nominal fee, has allowed him to walk out the door with a copy of the software. During his few days in Santa Fe, he has been experimenting with the Siemens neural net, seeing if it can predict Deutsche mark futures. Into his alchemical pot he has thrown spot rates, gold and oil prices, stock quotes, and interest rates. After propagating the data and pruning the weights, he has come up with what look like some "promising" models.

Finally, Norman gives a brief talk on "committee" models. This is a method for "training up different models and combining them in such a way that you get the best of all of them. The connectionist news group"—he is referring to the web site he visits for theoretical chats—"has been discussing how to 'vote' things together. As long as they're not correlated, you get a dramatic improvement in performance."

During lunch—a catered meal of Chinese noodles, chicken salad with snow peas, and Reine de Saba chocolate cake for dessert—the predictors quiz their guests. "What do *you* think moves the markets?"

"External news and internal psychological dynamics. These are the two things that drive the markets," Weinberger says. "The place we hope to find structure is in the internal psychological dynamics."

"Does one factor sometimes become more important than another?" Norman asks.

"U.S. Treasury bonds are news driven," says Weinberger. "Currencies less so. They're driven by psychological dynamics or big orders unrelated to news."

Suddenly the electricity goes out. For twenty minutes northern New Mexico is in the dark. This is a reminder that telecommuting to the world markets from Santa Fe is still a logistical gamble.

Weinberger pulls the change out of his pocket and starts flipping coins from hand to hand. "Where are we going to find structure?" he asks, rhetorically. "We're going to have to back into it. It's not going to come from first principles. And once we find it, we're still not going to be able to explain where it comes from."

"There are a huge number of factors moving the markets," says William, swallowing a piece of chocolate cake. "We'll never be able to identify all of them or weigh their importance."

"But some of this has to be figured out, or we can't trust our models. Upside outliers can be as scary as downside variance," Weinberger warns. *Upside outliers* refers to making money. *Downside variance* refers to losing it. Either prospect is troublesome for a quant, if it arrives unexpectedly.

The researchers keep peppering Weinberger with questions. They are curious to hear what professional traders think are the important numbers to use in predicting market swings. "We've never tried to predict direction," Weinberger confesses. "This is a much harder philosophical problem than the stuff we normally consider. It requires a depth of discussion we've never engaged in"

"You guys are asking the big questions," says Struve, joining William in a second helping of cake. He reminds them that they are following in the footsteps of other scientists who have tried to predict the markets: Isaac Newton, Josiah Willard Gibbs, Claude Shannon, Benoit Mandelbrot, Edward Thorp. Most of them proved to be better scientists than stock pickers.

"You are embarked on something like a car rally," says Weinberger. "You have to average a certain speed as you run past hidden check points. All the time you're following obscure clues. My favorite is, 'Don't split the match stick posts.' This refers to the blue and red-tipped snow posts on eastern roads. How do you pass them without 'splitting' or going between them? You turn left.

"Half the time, when you're looking for clues, you don't know whether they're up ahead or you just missed them or you executed the last command wrong, in which case the farther you go, the more hopelessly lost you're going to get. We face similar dynamics every day in our trades. When we're losing money, we keep asking ourselves, is it a statistical fluctuation? Is the structure of the markets changing? Are our models wrong?"

After a morning devoted to theory, the afternoon turns to practice. Up on the overhead projector come the "No Banana" and other graphs, which show Prediction Company's models performing well on historical data and then falling apart as they move into the present. "We may be having some problems with nonstationarity," Norman confesses. "It's not just slippage and commissions—although they hurt—it might also be model decay."

"What's this flat spot in the middle of your diagram?" Struve asks Stephen, who is projecting a graph labeled "The Death of jy.joe81b." This is a model for predicting the Japanese yen. For a decade, from 1980 to 1990, the model sails along with an annualized return of twenty-seven percent. Then, after 1990 it takes a dive.

"It loses money for two consecutive years," explains Stephen, pointing to the flat spot. "This shouldn't happen."

"Disappointed, were we?" jokes the old O'Connor researcher.

"You're into the bad news here," says Weinberger, pointing to three more graphs which show the performance of models for predicting the Japanese yen, the British pound, and the German mark. "Through the chaos of 1992 you were relatively stable. Your models hung in there. Then in 1993 they flipped."

"That's when you had German reunification," offers William. "There was a fundamental change in the European currency markets."

"I'm still puzzled by the anomaly of your models thriving on chaos and then falling apart as soon as the markets start acting normal again," says Weinberger.

"We could call a currency trader and ask what happened," someone suggests.

"That's dangerous," Weinberger counters. "It's like reading *The Wall Street Journal* to find out why the market moved."

"I take the Ecclesiastes view in these matters," remarks the Swiss

Bank equities man. "Every day is the dawning of a new age in the business press, but there isn't that much new under the sun."

"How do you know when epochs have changed so dramatically that you can't predict across the boundary?" Weinberger asks. "What George Soros did to the British pound wasn't an epoch change. The Gulf War wasn't an epoch change. These come along very seldom."

"It happens when you go off the gold standard or someone declares nuclear war," says Doyne.

"Not even that's certain to effect the markets," Struve says, making everyone laugh. "We might just keep on trading."

Weinberger parades around the table, jangling the change in his pocket. "Where do we go from here?"

Joe fingers the graph showing his failed attempt to predict the Japanese yen. "We've been conservative, holding back a lot of data for training our models. Maybe we should shift the window forward and go with shorter test periods."

Weinberger looks at him. "What's the shortest period to do reasonable testing?"

"It depends," says Joe.

"In turbulent times like these, I don't think you'll find a model that's good for more than a short time," William continues. "Then you have to go in and rebuild it."

"Don't throw these models out," advises Weinberger. "You might want to capture the decay process and pick out the factors that are causing it. All of us are groping with nonstationarity."

Weinberger is hanging on to his seat doing back stretches. Norman paces behind him. Joe and Stephen stand against the bookshelves. McGill is down on his knees, reinstalling a leg that fell off his chair. The meeting is beginning to hop.

"It's a real leap of faith," says Weinberger. "But I'm suggesting we build the best models we can on recent data."

"But what are the odds," Struve cautions, "if we produce a winning system, that we got our results at random?"

"You can figure this out," offers Doyne. "Even if you're flying blind on the internal workings of the model, there are statistical measures for this kind of thing."

"When you do adaptive modeling up to the present," counters Norman, "you burn all your data behind you. Without a test set, there's always the danger of overfitting the data."

"In terms of a pure experiment," admits Weinberger, "we're violating the rules. But everything we're doing here is a violation of the rules. We're not building an explanatory model. We're trying to get useful information. Even if we're not being mathematically pure, we can still be intellectually honest."

The room breaks into a free-for-all discussion of purity versus practicality. "How are you going to bet the ranch when the rules are constantly changing on you?" Heimark asks.

"The contract doesn't say Prediction Company is required to bet the ranch," Weinberger cautions.

"Just the upper forty," quips Doyne. Then he looks around the table. "So we're not really talking about betting the ranch?"

"Your contract was scaled originally to the O'Connor ranch," Weinberger explains. "Now that we have the Swiss Bank ranch, there is no way you'll ever come close to betting it."

"Forty-five percent of the time we'll be losing money, and every day we lose money I'll be nervous," Doyne admits. He is referring to the fact that even a winning system does not win every day. It wins on more days than it loses, but on its ride to victory, it can be expected to suffer some hair-raising tumbles.

Weinberger agrees. "You are going to confront a lot of gut issues as you watch yourselves lose money," he says. "But you people thought you could find structure. That's why we're all here. So the next step is to get into the markets and see what happens. Going live is the only way you're going to find out which parts of your models are robust."

Weinberger steps to the head of the table and delivers a pep talk. "A lot of people have found stuff on paper, but that's worlds apart from going out and making money. There's a huge gap between signals and real trading, where you measure profit and loss at the end of the day. Elegance is at the heart of this. I know you want to be pure. But now is the time to do whatever you need to do to get over the top. It's time to cross the chasm into the markets."

Staring down on the proceedings, with raised eyebrows, is the pop-art portrait of Albert Einstein. He looks bemused by the scene in front of him. There is the hint of a grin on his face as he sucks his pipe and blue smoke curls into his wild mane of flyaway hair. Printed on the bottom of the poster is his famous equation $e = mc^2$. It looks like a challenge from the grand master. "This is *my* guess at predicting the equivalence of energy and mass," he is saying. "Can you find anything this good for predicting Deutsche mark futures?"

Kinderkrankheiten

It is unlikely that a mere repetition of the tricks which served us so well in physics will do for the social phenomena, too.

—JOHN VON NEUMANN

Cecily, you will read your Political Economy in my absence. The chapter on the Fall of the Rupee you may omit. It is somewhat sensational. Even these metallic problems have their melodramatic side.

—OSCAR WILDE

Rafael deNoyo has made another bad business decision. When Doyne and Norman run into him again in October 1993, during one of their quarterly visits to Chicago, where deNoyo has recently relocated, they find him rich, successful, and miserable. He is being financed by a Christian commune that mixes chaos theory with God. He still subscribes to *High Times* and imbibes copiously, but the lack of friends is getting him down. The Caribbean vacation brochures have returned to his desk. It looks as if he will soon be taking another trip to the Bahamas.

When Norman last saw deNoyo on his porch in Richmond, Texas, in the spring of 1992, the muscle-man-turned-speculator was in the midst of being sued by Texas entrepreneur Ed Bosarge. While that diversion was getting sorted out, deNoyo, trading his own money, claimed to be getting returns of twenty-five percent, with a risk-to-return ratio that made his system look like a sure bet. This attracted two suitors to his door, who engaged in a bidding war for deNoyo's affection.

One was Michael Marcus, a former trader at the New York Cotton Exchange, who went on to become one of the world's biggest currency speculators. He made a fortune in gold and another fortune in cocoa before moving into trading tanker rates and other indices in the shipping industry. He parlayed a thirty-thousand-dollar stake into an eighty-million-dollar fortune. He owned ten houses in "every beautiful place in the world," many of which he had never slept in. His wife left him, but Marcus was too busy to notice. Trading from a beachside mansion in California, he was waking up every two hours throughout the night to place three-hundred-million-dollar bets on currency markets in Australia, Hong Kong, Zurich, and London. His secret? Marcus is a chartist. He is a trend follower who keeps an eye on market penetration and resistance.

The other suitor, who flew down to Texas in his private jet, was Joe Ritchie. Graduate of an obscure West Coast Bible college, he had been employed as a prison guard and policeman before going to work for a silver coin dealer in Los Angeles. There he got the idea that money could be made by trading on the small price differences that exist between the silver markets in Chicago and New York. He borrowed a threadbare suit from his brother Mark, who was also a divinity school student and night-shift prison guard. They made an appointment to talk to someone in the president's office at the Chicago Board of Trade. "You boys have got to be in the wrong place," he guffawed, when they walked into his office.

The Ritchie brothers began doing silver arbitrage in 1970. They switched to soybeans and then moved from there into the T-bond futures options pit. "In 1975 I crammed the Black-Scholes formula into a TI-52 handheld calculator, which was capable of giving me one option price in about thirteen seconds," remembers Joe. "It was pretty crude, but in the land of the blind, I was the guy with one eye."

In 1982 the Ritchie brothers jumped into the new Merc pit for trading stock index futures on the S&P 500. As more new markets opened in London, Paris, and Singapore, they expanded into trading currency futures and options. Within a decade, their company, Chicago Research and Trading or CRT, had grown into an eight-hundred-person firm with an estimated one billion dollars in profit.

Trading in seventy-five markets on nineteen exchanges around the world, CRT was only slightly less sophisticated than Chicago's other powerhouse dealer in derivatives, O'Connor & Associates.

In 1993, when America's seven largest banks—the so-called money center banks—wanted to goose up their profits by playing with derivatives, the Ritchie brothers sold CRT to NationsBank for two hundred and twenty-five million dollars. Explaining why they sold out, Joe says, "We used to pick up dimes in front of bulldozers. Then we were picking up nickels and pennies."

Mark Ritchie penned his memoirs, *God in the Pits,* a blend of spiritual revelations and trading lore. Joe Ritchie founded a born-again Christian commune in Fox River, Illinois, forty miles west of Chicago. Then he bought Rafael deNoyo's company for two million dollars and set out to make his second fortune.

After dinner at an Italian restaurant in Chicago, deNoyo drives Doyne and Norman out to a big house in the middle of cornfields. Inside is a party of Baptists—"the cult," deNoyo calls them—who are swapping stories about their adventures as missionaries in South America. They seem to have spent most of their time on hunting trips shooting snakes and monkeys. Someone picks up a Bible and starts reading an Old Testament curse. "This is for Bill Clinton," he declares. Doyne and the Bible thumper are headed for an altercation, when deNoyo takes his guests by the elbow and steers them next door to the farmhouse that doubles as his research center and global trading station.

The house is crammed full of computers being tended by one of deNoyo's colleagues from Texas. Being delivered into his living room by satellite dish and T1 phone line is a big pipe load of tick data on currencies, stock indexes, and interest rates. DeNoyo spends eighty hours a week cleaning up the data feeds, massaging the numbers into meaningful patterns, and following the book as it moves around the globe. He is playing twelve international markets, everything from the Nikkei to the Footsie. He breaks out the wine and cigars and settles down for an evening of talking shop and politics and other topics covered in *High Times*, his favorite magazine. "It's the only place where you can learn the truth about a lot of stuff that's going on," he remarks. He is making good money, but the hours are killing him, and the Christian scene is getting a bit heavy. So he is thinking of establishing an 800 number in a foreign country and becoming a bookie. "Sports betting and online gambling, there's a lot of money to be made there," he declares, warming to the topic, as he opens another bottle of wine.

The last three months of 1993 are make-it-or-break-it time at Prediction Company. As everyone jams to get a suite of winning models online, the company turns into a hive of number crunchers. "We trade everything," says Clay Struve of the derivatives contracts he handles in Chicago, "so long as it doesn't walk, shit, or rot." In other words, he does gold, but not cattle. Crude oil, but not wheat. With this as their only limitation, Prediction Company begins moving into every one of the world's potentially beatable markets.

Behind the innocuous door with "Prediction Company" written in small letters is an alcove where Laura Barela sits at a tidy desk. She is the cheery, freckle-faced woman who manages the office, pays the bills, and keeps track of Prediction Company's ever-mounting collection of research reports. She is the calm presence that more than anyone makes the company look like a company.

Beyond Laura the room opens into the big skylit space which is filled with the "board room" table, library, potted plants, couch, and trading console. The first office to the left of the front door, a room with grape leaves trailing across the exposed struts in the ceiling, is occupied by Joe and Stephen. Apart from two whiteboards covered with equations, the room is decorated with a big poster, TRADER'S GUIDE TO THE WORLD, which lists the globe's sixty-nine futures markets, from Argentina to Singapore, and everything they trade, from stock indexes to raw silk.

Today's problem is the gold market. Joe and Stephen have found lots of structure in the data coming out of COMEX, the Commodity Exchange in New York. These are patterns—like tells for a poker player—which indicate which way the market is likely to move. But there is also something fishy about the data. They display a wide variety of statistical aberrations. These imply that the benign hand of the market is being tinkered with by other hands not so benign.

"This is interesting," announces Stephen, who is leafing through a book on gold trading published by the International Monetary Fund. "Do you know why it's called the 'daily fixing' when they announce the price of gold in Zurich or London? Because the price is 'fixed' by something called the 'gold pool.' The closing price is not the price at which the last gold contract changed hands on the open market. It is the price established by a cartel of bankers meeting behind closed doors."

"No wonder the time series for gold is squirrelly," Joe remarks. "Maybe we should try nickel or copper."

"They don't have the volume," Stephen replies. "When it comes to metals, gold is the big one."

"So we won't trade the closing price," Joe concludes. "Unless we can get admitted to the cartel doing the 'fixing.' "

The next office along the east wall is Norman's. He sits at an enormous L-shaped desk covered with piles of books and papers and two computer terminals with nested screens running e-mail, batch processing, neural nets, a game of Go, and the promising early stages of a Japanese bond model. Norman's office has two big windows. One facing out to the Santa Fe mountains. The other facing *in* to the company's big central atrium. This is the kind of window a manager uses to monitor the company time clock. But Norman is famous for keeping his eye on no clock, save for the internal rhythm that allows him to work through the night when he is chasing ideas or scrounging CPUs from remote computers.

Next in line is Doyne's office. The walls are decorated with shots of earth from outer space, children's drawings, and a stereo picture of the Chaos Cabal taken with Jim Crutchfield's 3-D camera. Doyne works at a wooden table facing a window that looks out through a stand of cottonwoods onto the state capitol. Behind him is another table and chairs and a sofa covered with an old madras bedspread. He keeps the blinds pulled against the glare, and the absence of sunlight might explain why the hanging asparagus fern is shedding onto his computer. The room feels warm and clubby, a feeling enhanced by the steady stream of people who walk in during the day to talk about model building, test results, bugs, glitches, and the Footsie, which is Doyne's latest research obsession. Open on his desk is one of the notebooks in which he has been scribbling equations and bright ideas ever since he first acquired the habit in Silver City.

The last eastside office is Jim McGill's. The walls are bare, the space so expressionless that dusting it for fingerprints might produce scant results. In fact, he continues to spend much of his time on the road. Swiss Bank is effectively paying half his salary, while McGill works for them as a consultant evaluating financial data services and designing the pipeline that will deliver them to the desktop of every Swiss Bank trader. There are only two objects in the office that hint at McGill's personal life. When he is residence, he places them next to his PowerBook. One is a two-thousand-year-old Colombian fertility goddess, a little red stone figure of obvious fecundity. The other is a picture of his Thai girlfriend, whom McGill is about to make his second wife.

The south side of Prediction Company's second-story loft is given over to the bathroom and Jacuzzi, the kitchen, which is stocked with everything from granola bars to champagne, and the old darkroom, which is now stuffed with computers, routers, modems, tape drives, local area networks, wide area networks, and other gear required for sucking numbers out of the ether and transforming them into predictions. This is the domain of Paul Ford, a ponytailed New Mexican who has been hired as Prediction Company's systems administrator. Ford is the company's one true geek. He consumes great quantities of sci-fi and junk food. His smudged black plastic glasses are constantly slipping off his nose. He is a *Star Trek* couch potato in a company of mountain biking sports freaks. But he keeps their machines running, which is no easy task, given the polyglot assortment of equipment, ranging from homemade to high-end. The pieces are wired together into one big computer that is continually being expanded to satisfy the company's voracious need for central processing units.

On its western side Prediction Company's loft has a single office shared by Nusbaum and Finnoff. Taped on the wall next to Jim's desk are maps of Taos, Vail, and other Rocky Mountain ski resorts. Lately, he has been too busy for more than the occasional run down Santa Fe Baldy, where he likes to ski off-trail, substituting tree trunks for slalom poles. This is fun when it works, but last week he got hugged by a fir tree and came away with a black eye.

Jim is busy dealing with the information technology people at Swiss Bank, who are having a hard time figuring out how to move some of Prediction Company's machines to Chicago. After a few visits to LaSalle Street, he begins to get a reputation for being an exacting critic. But he is also a computational genius, and no one can argue with success. For Nusbaum, combining their two systems would not be worth the trouble if Swiss Bank Corporation did not have one thing Prediction Company needs: data. They have a big pipe, called the Price Machine, which supplies their traders with real-time data from most of the major data suppliers known to *homo economicus*.

The financial data business is changing so fast that new suppliers pop up every day. A lot of ticker-tape data that used to fall on the floor is now being picked up by companies that sell it for hundreds of thousands of dollars a year. As the markets become increasingly mathematized, and as more and more computers are enlisted to analyze and run them, the numbers generated by these markets become increasingly valuable. For Prediction Company to emerge as the world leader in particle finance—which is the true ambition of these otherwise laid-back New Mexicans—it needs the numbers, all the numbers, which Nusbaum is struggling to push westward in a data pipeline longer than the old Santa Fe Trail.

Finnoff builds predictive models the way Nusbaum skis. He runs at the problem nose first, moves fast over unmarked terrain, and keeps pushing himself to crank out as many models as possible. They don't have to be perfect. Just swift and competent. He, like everyone else on Aztec Street, feels impending doom breathing down their necks if they can't get a suite of winning models in place by the end of the year. He burns the midnight oil. He works weekends. He ignores his girlfriends and spends long hours staring at a computer screen whose color he describes as "baby-shit brown."

From Siemens he licensed and brought with him a neural network with a hundred thousand lines of code. This is the Mercedes Benz of neural nets, engineered with German precision and solidity. The program produces diagrams of what look like spindle-legged spaceships stacked on top of one another. In this Tinkertoy arrangement of pods and connectors, the two pods at the base of the diagram represent input nodes. The lines connecting them represent weights. The pods hovering above represent hidden clusters. The top pod is the output cluster. This is a feed-forward network capable of massaging hundreds of inputs with thousands of hidden neurons. The pods, burnt sienna in color, are overlaid on top of a plum red background. Their activity as they start to "learn" is measured by an array of cobalt blue gas gauges. The look of the system is strong, manly, Germanic. For someone seeking "decision support," which is what it was designed for, the network's mystical workings are cloaked in just the right note of sober assurance.

In less than a month William has used his program to build eight "quick and dirty" models. Two of them look good enough to bet on. One predicts the price of Treasury bond futures. The other tracks the S&P 500. As for testing the validity of these models, "A lot depends on personal experience," he claims. "You can't patent it." His models might be good, or they might be dumb monkeys who have learned nothing more than how to predict the past. One way to find out is to put them into production and see if they make money.

While William throws himself into new data sets, the software programmers are left with the task of putting his models online. This provokes the usual crises that dog Prediction Company every time they move from theory to practice. "It's like building a race car," says William, whose father used to be in the business. "You don't get to test enough, and a wheel falls off. So you reassemble the package and wing it. Fortunately for the company, the software group is really good. They're the ones who are keeping us in the race."

Another concept imported by William is described by the German word *Kinderkrankheiten*. "Glitches with the software, the screwup where we forgot to take account of daylight savings time, our problems with rollovers and holidays—these are nothing more than

Kinderkrankheiten, childhood diseases," he declares. "You get them with any new company. You also outgrow them."

William begins slashing his way down another difficult slope when he joins Doyne and Norman in trying to figure out how the oil market works. First of all, there is not one oil market, but several. West Texas intermediate light sweet crude is traded at NYMEX, the New York Mercantile Exchange. Brent-blend and unleaded gasoline are traded at the IPE, the International Petroleum Exchange of London. High sulphur fuel oil is traded at the SIMEX, the Singapore International Monetary Exchange. West Texas intermediate light sweet crude, known as WTI, is the third most heavily traded commodity in the world.

One futures contract equals a thousand barrels of crude oil, which typically sells for around twenty dollars a barrel. Ninety million barrels a day are traded at the NYMEX. Consider the fact that only sixty million barrels are produced each day, and one can see that speculation is the dominant component in oil trading. Much of this speculation is conducted by "Wall Street refineries," investment banks like Morgan Stanley and Bankers Trust. The trick is to keep these contracts rolling over before they expire—making a few pennies on the churn—because no Wall Street refiner actually expects to see a tanker pulling up to the front door.

Swiss Bank is new to this game, and Clay Struve, speaking with the voice of someone who still keeps the company's risk control position in his head, has only one word of advice to his colleagues in New Mexico: "If Swiss Bank ever takes delivery on an oil futures contract, we're all out of here."

On the north side of Prediction Company's big central room is a windowless office with skylights. The skylights are pleasant, except when it rains and the water leaks down on the company's software group, which now includes Jim Nusbaum and two other people. Doug Hahn is responsible for taking the sewer pipe of raw data that arrives daily at Prediction Company and cleaning it with mathematical solvents known as "transforms." These alchemical functions turn ticker-tape prices—the dross of global finance—into logs of prices or differences in prices between today and yesterday or the twenty-day trailing average of the price. Knowing how to apply these tricks of the trade is the first secret to building predictive models. Transforms are to particle finance what experimental design is to particle physics. It is the first step in turning disparate data streams into a time series, which is the golden thread out of which Prediction Company weaves its predictions.

Doyne compares Hahn's technical transformations to what the brain does when it recognizes faces. "There is heavy preprocessing in the brain, with edge detectors and motion detectors, before the inner brain processes information presented to it by the eye. If it didn't do this, the brain would be swamped by an undifferentiated mess of pixels."

Sitting on the other side of the drip bucket from Doug is Karen Lawrence, who arrived on Aztec Street in December 1993. She is charged with designing a new trade engine. She is a software engineer, hypercompetent and casual, who fits in immediately as one of Prediction Company's fun-loving polymaths. She is the trim mother of three, with bright blue eyes and long dark hair framing an even-featured face. She is a Texan, with an easy smile, who comes to work in blue jeans and hiking boots and disappears from the office at noon for a quick game of racquet ball. In the predominantly male world of software engineers, she wears her talents lightly.

Short of space, Prediction Company locates two more employees out the front door and down the corridor in a room called "Siberia." One is Jan Scanlan, the sunny blonde who has replaced Helen Lyons as the company's part-time accountant. The other is Sonia Fliri-Hummer. Sonia is a Tyrolean whose parents own a shoe store in the Italian Alps. She wears her dark hair short and nicely styled. She has lovely Italian knit sweaters and great shoes. She is pink-cheeked, young, cheery, and the company's only trained economist, with a master's degree from the University of Vienna. Without a green card and still in the early stages of adding English to the three languages she already speaks, she begins working at Prediction Company part-time, helping Doyne write his blockbuster "Summary of World Futures Markets." Soon she starts getting a salary. Then she is working full-time. Eventually Sonia emerges as the company's data queen. She knows everything one needs to know, or where to get it, about the world's markets and over-the-counter financial exchanges. If information about money is as important as money itself, then Sonia is worth millions.

"It's not really funny to do this work," she says, in her Italianate English, "but it's really important." It is so important that she regularly gets phone calls from headhunters in New York offering to triple her salary.

Once this team of thirteen predictors is in place, they throw themselves with renewed vigor into cracking the markets. They have until the end of January 1994 to get a winning system online. This is when Doyne and Norman are scheduled to meet the Swiss Bankers for their next quarterly meeting. They are hoping—if they can pull it off in time—to ask that SBC's Prediction Company portfolio be enlarged tenfold, with each of their models trading between ten and fifty contracts a day. This is not a big position for Swiss Bank, but it is a hell of a gamble for some old poker players from Silver City. "We have to keep cool," Doyne advises. "This could be a bumpy ride."

They embark on a big push to develop winning models. They write and debug thousands of lines of code. They transform copious streams of data. Everyone in the company rides the same emotional roller coaster. One day they are down because someone has discovered that the Footsie model, by accidentally looking across time zones, is contaminated with future data. The next day they are up because Karen's back tests show them making money hand over fist.

Four models are functioning by New Year's Day. Two models, using Prophet and genetic algorithms, predict price movements in the Japanese yen and the German mark. Two models using neural nets predict price movements in U.S. Treasury bonds and the S&P 500 stock index. Another model, almost ready to go into production, predicts the price of oil. Waiting in the wings are models for the British stock index, German and Japanese treasury bonds, and the gold market. The four operating models have passed their back tests. The trade engine is firing on all cylinders. The data pipeline into Aztec

Street is pumping at full volume. "We're ready for liftoff," Norman declares. "It's time to tell Swiss Bank to push the button."

On January 26, 1994, the Technology Development Committee, which is the official name for the group that oversees Swiss Bank's relationship with Prediction Company, records in its minutes that Doyne and Norman have provided "proof of principle" that their models work. Prediction Company is deemed to have crossed the contractual threshold that allows SBC to give them the green light to ramp their position.

Craig Heimark has flown in from Switzerland for the meeting. "Let's put the pedal to the metal," he says. He advises Doyne and Norman that Prediction Company "looks thin on operations. You should probably think about hiring more people. Once you put this much money in play, there's a lot of live trading to monitor."

If they are going to hire more people, Prediction Company needs an infusion of capital. They talk about a bridge loan of two hundred and fifty thousand dollars, to be converted eventually to stock. Heimark, Weinberger, and Struve, with the bank's approval, offer to put up the money themselves, with shares in the company valued around twenty-five dollars. This is a vote of confidence. It is also a last-ditch effort to keep Prediction Company flying with a low profile. Everyone in the room senses that they will soon be on the world's financial radars.

The Curse of
Dimensionality

Most of us enter the investment business for the same sanity-destroying reasons that a woman becomes a prostitute. It avoids the menace of hard work, is a group activity that requires little in the way of intellect, and is a practical means of making money for those with no special talent for anything else.

— RICHARD NEY

A mind is a horrible thing to invest.

— JIM CRUTCHFIELD

On a spring day in 1994, the weekly research meeting is delayed while Jim McGill talks to two technicians from New Mexico Satellite, who are installing a receiver near the company's rooftop gazebo. The satellite dish, roughly the size of a pizza tin, will be pointed at Spacenet 3, which is one of twenty-two satellites parked over the state, this one specialized in transmitting data.

The Santa Fe rooftops are dusted with snow. The mud-brown village smells of burning piñon and pine. The sky is black and blue from a spring storm, and up on Santa Fe Baldy a squall is raging. The clouds are scudding fast, beginning to break up, and off to the west, over the Rio Grande, a shaft of golden light beams down on mesas that are glowing like freshly baked buttermilk biscuits.

"We think we got it pointed in the right direction," announces César, the chief satellite installer. "We put up a lot of big dishes, you know, for TV. But we don't know much about these little ones."

"The problem with these small dishes is that they don't work too good when it snows," says César's companion.

"What do you mean?" asks McGill.

"The snow gets in the dish and affects your transmission."

"Then what do you do?" McGill's pink face begins to darken into a monkish scowl.

"You go up on the roof and sweep off the snow."

"You mean to tell me that, if I'm trading Japanese yen in the middle of the night, and there's a storm, I'm supposed to send someone out on the roof to dust off the snow?"

"Yeah, I guess so."

The technicians have nothing better to suggest. So McGill gets on the phone to Knight-Ridder, the company supplying the data being beamed down on Aztec Street. His blue eyes go cold as steel. "You guys are selling a twenty-four-hour data feed to people trading in the global markets, and you're telling me that everyone who subscribes to your service outside of California and Florida is supposed to stand around with a broom in his hand?"

Getting nowhere, McGill phones the company that manufactures the dish. Yes, snow is a problem, they admit, but one solution is to buy a clear plastic snow dome that snaps over the receiver. "Send me one, express mail," orders McGill, muttering about the impossibility of doing business in New Mexico.

McGill is not the only person having a bad day. It is a sober group of researchers who are gathered around the boardroom table. After the push to get more money behind their models, the models are looking beleaguered. They are not performing as hoped, and there is a laundry list of problems to be confronted at this glum Monday meeting.

William, as if girded for battle, is wearing blue jeans and a black belt with a big silver buckle. Joe and Stephen are dressed in blue work shirts with their sleeves rolled up. Doyne is wearing Patagonia climbing pants with extra padding at the knees. McGill's pink-rimmed eyes are staring warily from behind his tortoiseshell glasses. Norman sits cross-legged in his chair, his long face framed by the helmet of blond hair that drapes over his shirt collar. Jim Nusbaum is dressed with

Lutheran severity in black chinos and a gray knit sweater; even his ponytail is tied up with a black-and-white hairband. Karen, Sonia, and Doug sit back from the table, as if trying to stay out of the line of fire.

The first item tackled at today's meeting is the "agreement" problem. Agreement is the gap between predicted results and actual results. "This one problem is really twenty problems," Doyne declares. "It's not on anyone's quarterly objectives because it came out of left field. But we're sunk if we can't get a handle on it."

The problem afflicts models that are scheduled to fire, but are delayed by missing data or confused by spurious data, which is transmitted from the markets and then later "corrected." Jim reports that he's been learning "some weird stuff about the time stamps on our tick data," the record of price movements in markets. These data are broadcast in strings of numbers with time stamps, which indicate that IBM traded downward at 3:04 and then traded downward again at 3:05. He has discovered that the original numbers put on the data by the New York Stock Exchange are altered by the private companies over whose lines this data is broadcast.

"Most vendors 'restamp' exchange data at their ticker plant. So you seldom get the numbers coming off the floor. The other problem is resolution," he says, referring to how often the data are stamped. "Reuters and Telerate have a sixty-second resolution. Telekurs has a thirty-second resolution. Tick Data Incorporated has the finest granularity, one-third of a second, but all the data they originally sent was corrupt."

"What a piece of rubbish!" Doyne exclaims. "Why don't they keep the original time stamps? If I were collecting tick data, I sure as hell would report it the way it comes off the exchange."

"What if we run in a dedicated phone line and start collecting our own data?" Norman suggests.

"We're dreaming here," says Doyne. "The bill would be in the hundreds of thousands of dollars."

Nusbaum goes on to the next problem, volatility. "When the markets get busy, the data coming over the wire lag further and further behind the actual events. For our applications, late data are sometimes worse than no data at all."

The satellite guys interrupt to say they are going to look for a hundred feet of missing Teflon cable, which is required to get their system operating. Assuming they finish the job, Prediction Company will be flooded with numbers from Telekurs, Knight-Ridder, DRI-McGraw-Hill, Datastream, TDI, Dow Jones/Telerate, Market Vision, and a host of other data providers. "Someone is going to have to take responsibility for getting these people to talk to each other," Nusbaum warns. "At the moment their systems are completely incompatible."

McGill, as consultant to the Swiss Bank Data Project, along with everyone else at Prediction Company is in hot pursuit of what they call a "real-time ticker." This is a live data feed straight out of the markets. Onto this data is clamped a battery of monitors for reading the financial pulse of the world. "Even if we could get into the tick stream in real time, do we have the software to do it?" asks McGill.

Norman is optimistic. "The stuff from Datastream looks pretty good," he says. "But they charge forty thousand dollars for their bond data alone, and that's more than we pay for all the rest of our data combined."

"Maybe SBC will pick up the tab," McGill offers. "But before we ask them to pay for it, somebody should take a closer look to see if it's any good."

"Sometimes the DRI data doesn't come at all," Nusbaum continues. "How do we factor this into our models?"

Doyne scribbles a note to himself. "It has to be done model by model. It's a time sink, but somebody has to be in charge of getting back into the models and cleaning up the data." Then he brings up another item on the laundry list. "We need a flag in the database telling us when there is something wrong with the feed."

Nusbaum rocks back in his chair. "These flags add a lot of baggage."

Doyne nods in agreement. "I know. But if we indiscriminately drop all the corrupt data, we could end up with *no* data."

He moves on to the next item on his list. "What I really want to do is build an incidence table that compares every 1 and 0 in all the data sets and live feeds."

"There are five possible back tests for every model," Stephen cautions. "It's not like we can dive into it and say, 'Oh, we have this sign wrong.' " Then, with an air of resigned skepticism, "It's much more complicated," he says. "There is so much data to reconcile, and you can reconcile in so many different ways that it's hard to say exactly where things are going amiss. The data are mostly right. The transforms are only slightly less reliable. The predictions are off, and the fills are totally off. The errors propagate downstream, growing exponentially."

Ultimately, as Prediction Company deals with increasing complexity in an expanding number of markets, they run up against the curse of dimensionality, which plagues systems with too many variables to comprehend. The predictive power of their models lies in making patterns appear in an imaginary realm called state space. A *state* is the mathematical description of a system, such as a dripping faucet or the stock market, at a particular point in time. The *space* is where changes to this system are plotted graphically.

State space diagrams, which allow physicists to think about the world in geometrical terms, are an enormously powerful tool. The dimensions in state space correspond to the variables that describe the system. For a pendulum, the state space is two-dimensional. One dimension represents the position of the bob, the other its velocity. For a dynamical system with three variables, the state space is a point moving through three-dimensional space. A state space might have an infinite number of dimensions, but in this case, one would need an infinite amount of computing power to extract any useful information from it. One would also need an infinite amount of data since, as a general rule, there is exponential growth in the number of measurements required as the number of dimensions expands.

State space reconstructions, which were first used by the Chaos Cabal back in Santa Cruz, help scientists discover if deterministic patterns exist in irregular data. They succeed only if the data are hiding deterministic patterns in the first place, and they are most useful when they involve a limited number of dimensions. The *system* being reconstructed may have lots of dimensions, but its *attractor*—the geometric shape around which it travels through state space—should have a limited number. The stock market is a mess of variables, to be sure, but a mere handful of these variables might describe its attractor, or stranger-than-strange attractor, as the case may be.

The advantage to thinking in terms of state space is that no matter how many variables are involved, the system is represented by a single point. As the dynamical system evolves, the point zipping around in state space reveals by its position exactly how the system is changing in time. "We take the numbers and create space for them to tell their story," Norman says.

The difference between systems that are predictable and systems that are not predictable lies in the number of degrees of freedom they possess. A roulette wheel has many degrees of freedom, but roulette wheels are, to some extent, predictable; the stock market may have no more randomizing mechanisms built into it than a roulette wheel. Once the croupier releases it, the roulette ball tumbles through a mechanical universe designed to scatter it. But stock prices are never released from human influence, and humans are herd animals who tend to follow relatively few fads and fashions. Investor preferences, trading lore, and human foibles appear to be one main source of both the chaos and pattern in financial markets.

Harry Seldon, the hero of Isaac Asimov's *Foundation* novels, invents something called the "science of mobs," which allows him to predict thirty thousand years of future events with "something of the accuracy that a lesser science could bring to the forecast of the rebound of a billiard ball." Norman, whose computer is called Seldon, believes the science of mobs would be quite useful for predicting the stock market. "There's nothing mystical about it," he says. "It's just a huge pile of numbers."

But eventually there is a limit to what Prediction Company's models can handle. No matter how sophisticated their techniques, they ultimately bump against the curse of dimensionality. "Just adding more dimensions may give you more room to fool yourself," Norman warns. "What you want is a minimal network strategy. You want to bound the complexity of the model and go with the simplest one available." Elegance of design rather than brute number crunching is the way to solve this particular problem.

Corrupt data, nonstationarity, stranger-than-strange attractors, the curse of dimensionality—legion are the problems weighing down this sodden Monday meeting. Another problem is the fact that Stephen and Joe are increasingly at odds with everyone else about where to look for solutions. Stephen, an expert in time series analysis and genetic algorithms, had had some early success applying these methods to the currency markets. But since then, breakthroughs have been scarce. Jim McGill is trying to push him into using neural nets, but Stephen does not believe in neural nets. They are a computer-intensive fad, he thinks. They are too prone to reading meaning into mush, and Joe agrees.

Joe used to live in a simpler world. His dad ran a goldfish hatchery; his mom ran a bridal shop. Born in 1965, just as the Vietnam War was beginning to monopolize the evening news, he was a bright kid who studied physics and math at the University of Indiana, then crossed the border to go to graduate school at the University of Illinois. He worked for five years in the nuclear physics lab at Indiana. He loved the experiments, the hands-on research, the cutting and grinding involved in making atomic detectors. He became an expert machinist and experimental physicist, but nuclear physics is a dying field. So he moved into astronomy, which was another childhood romance; here again the field was changing. It was getting away from star gazing and becoming more abstract. He was advised to import chaos theory into his study of globular star clusters. He sought out Norman at the Center for Complex Systems Research and asked if he could work with him. Norman put Joe on the payroll, then disappeared to Italy for eight months.

Joe, on finishing a doctoral dissertation in nonlinear dynamics, again found job hunting difficult. This time the problem was not a dying field, but a field that has never really been born, except as the stepchild of the various disciplines it straddles. Like many of today's newly trained physicists, the best place Joe could think to look for work was Wall Street. He told Norman he was converting his techniques for analyzing star clusters into financial forecasting tools. This is when Norman invited him to Santa Fe.

From the moment he walked in the door, Joe could never shake his moral qualms. His scientific heroes smashed atoms and stared through telescopes into far-off galaxies. But no atom smashers or telescopes are involved in financial forecasting, and he kept worrying the question "Who are we helping here? As far as I can tell, all we're doing is taking money away from widows and orphans."

Joe is not the only person at Prediction Company pondering the morality of beating the markets. One day, as Norman logs into his connectionist newsgroup, he finds an electronic message posted from "Bill" at the University of Arizona. "Motivated by the announcement of the 'First International Nonlinear Financial Forecasting Competition,' I would like to raise a concern that has troubled me for some time," writes Bill. Norman and Doyne are helping to organize this competitive test of financial forecasting models, and Doyne is serving as a judge.

"I wonder whether it is really socially responsible to work on these things, or to support such work," Bill writes. "The financial forecasting techniques I have seen are not oriented toward predicting the failure or success of individual enterprises, but rather toward identifying and predicting global trends in the flow of money. I am skeptical that this is possible in the first place, but even if it is possible, it seems to me that to make money this way is to be a parasite upon the financial system rather than to serve it. Furthermore, it is pretty clear that the only way to make money consistently with such a technique is to keep it secret."

Norman takes a stab at addressing Bill's concerns. "Finding predictable structure and trading on it makes the markets more efficient," he replies. "Such trades reduce the overall volatility of the markets, and thus the risk to participants is reduced. The profits that go to the successful predictor are like a fee that the markets pay for the service of making them more efficient and less risky."

According to Norman's argument, financial forecasters, far from being parasites freeloading off the labor of others, are the good guys of capitalism. They help to keep the markets deep, liquid, and transparent, which are economic terms that were borrowed from physics in the first place. *Deep* means the markets are full of money. *Liquid* means the money is flowing from pocket to pocket. *Transparent* means the flow is not being blocked by inside traders, monopolists, price fixers, and other cheats. If nothing else, the financial models being developed by Prediction Company are the best fraud detectors currently available. They work like a kind of polygraph clamped on the economic pulse of the world. They are hypersensitive measures of monetary flows and volatility, which pick up market anomalies in a flash. If the lawmen at the Securities and Exchange Commission ever get their hands on this technology, it will be a regulator's dream.

Doyne takes another tack in defending the morality of what Prediction Company does. "The traders in the pits are part of the decision-making apparatus that sits in the collective brain of human culture. The prices at which commodities are traded dictate whether farmers grow more wheat or soybeans. The prices at which stocks are traded dictate whether capital investment is directed toward making cars or computers. Interactions among traders determine the prices and allocate the resources for all of society. Like democracy, it is a highly inefficient system that doesn't work very well. But it seems to be superior to the known alternatives, such as a centralized economy.

"Making predictions is fundamental to financial markets," he adds. "If people didn't think prices were going to change, they wouldn't trade. Price setting can be rational only insofar as the predictions that drive it are sensible. By making predictions and trading on them, we are indirectly helping society set its goals, even if we're also trying to make a profit."

His final thoughts on the subject are cautionary. "Whatever we say about the morality of financial trading is likely to be wrong," he warns. "Financial markets are self-organized systems whose emergent properties are not well understood. Speculators have been accused of making markets crash, ruining national economies, and other evil deeds. But without speculators the markets wouldn't work. I'm sure there's something better, but I don't have a clue what it is, and I'm wary of people who say they do."

The sky outside has cleared to eggshell blue, but inside the predictors are still sitting under a cloud. Norman's rash has reappeared, along with Doyne's insomnia. For the last few weeks they have been wondering what to do about Joe and Stephen's growing isolation. "If they decide to leave, it will be a clean sweep," says Norman, who is talking to Doyne while standing in the morning coffee queue at the Aztec Street Café. "All the original founders, except for you, me, and Jim, will be gone."

"It's the end of the Eudaemonic ideal," says Doyne. "Instead of a shared vision, we'll just be another company."

Across the street, where he is busy unpacking the clear plastic snow dome, which has been sent to protect the company's satellite dish, McGill is philosophical about the latest turn of events. "This whole experience is about the education of Doyne and Norman," he says. "That's what the last three years of work and all this investment have been about. They are the franchise players. All I can do is provide them with the tools they need: money, time, software support, knowledgeable business partners. They can't say they haven't had a run at the wall. Has it made them better people?" he asks rhetorically. "Hell no! But they've learned a lot about money and how to make it."

While Doyne and Norman suffer sleepless nights and fret about what to do, Joe provokes the final decision. One day he walks into Jim McGill's office and says he wants to sell his Prediction Company stock. For a founder to sell out before a company is successfully launched is a vote of no confidence. What is Joe going to do after selling out? Put his feet on the desk and read the newspaper?

By the end of June, Joe and Stephen have decided to walk out the door to new jobs. "At some point, graduate students establish their independence by killing their father, i.e., their adviser," Norman realizes. "This is healthy for the development of a young scientist, but it doesn't work in a start-up company."

Doyne offers another analogy. "It's like getting a divorce," he says.

"It doesn't really matter whether you were right or wrong or who was the bigger asshole. In the end all you have is a confession that something you started out to do has failed."

On August 9, 1994, with oil, currency, and futures models firing faultlessly in a rebuilt trade engine, Prediction Company flips the switch. This is the big bet, the tenfold ramp approved by Swiss Bank at the January meeting. They are making the move from low-level testing to real money. Signals go out to trade millions of dollars in futures contracts. A big cheer goes up among the predictors crowded around the trading console. It has been a long haul, over a bumpy road, and now that they are nearing the end of the trail, they are ready for a victory lap around the plaza.

A few days later, as soon as Doyne and his dog walk in the front door, they sense that something is wrong. Clara omits her usual *woof*. The crowd gathered around the trading console is wearing long faces. Doyne strides toward the monitor. He can tell at a glance what's going on. "We're getting creamed!" he shouts from halfway across the room.

"The Fed is adjusting short-term interest rates, and the bond markets are going crazy," says William. "It was a surprise announcement."

The monitor offers a brightly colored display of what is happening tick-by-tick in Chicago, New York, and other major markets. Yesterday the trend lines were tilted nicely upward. They augured clear sailing through light seas. But today the markets are plummeting so fast that the green trend lines are beginning to break apart as they fall relentlessly toward the bottom of the screen.

Next to the spinning numbers and gapping trend lines is another screen showing SBC's Prediction Company portfolio. This lists contracts, fills, orders, and daily profit and loss. Today the P&L is in brackets. This is how accountants record the L part of P&L. "The market always overreacts," cautions McGill. "Don't be surprised if you see it turn around as fast as it's dropping."

By the end of the day, the crowd huddled around the trading console have watched the portfolio lose more than a hundred thousand dollars. By the end of the week, it will have lost a good bit more.

To steady people's nerves, Norman tapes to the wall above the trading console a graph showing Prediction Company's probability of taking losses of different sizes. The graph illustrates a gambler's chances of winning or losing x dollars in a game played at y odds. Even a gambler playing with a large advantage stands a remote chance of being wiped out. In the long run the numbers may be on your side, but in the short run, statistical fluctuations can scare the willies out of you, or if your betting pool is insufficient to the task, they can drive you into bankruptcy.

To counter this risk, Prediction Company is relying on something called the law of large numbers. With one coin flip, there is a fifty-fifty chance of getting heads. With ten coin flips, the chance that all ten coins will come up heads is roughly one in one thousand. Similarly, if ten Prediction Company models are trading in ten different markets, the chance of them all losing at once is much smaller than the chance of any one of them losing.

The odds are on their side. SBC's investment capital is ample. Their potential gain is handsome. But still, about once a month, they should expect to lose a hundred and fifty thousand dollars in a single day. Of course, they plan to have more winning days than losing days, and it is bad luck that they hit a Fed announcement so soon. But people looking at Prediction Company's P&L are going to have to get used to seeing large numbers—with or without brackets around them.

Zozobra is burned at Fort Marcy Park the weekend after Labor Day. He is gloomier than usual this year, pallid with worries, and his down-turned mouth adds a note of misery to the puppet's scowl. It is a beautiful night for the opening of Fiesta, star-filled and balmy, as most of Prediction Company gathers on the lawn for their yearly picnic.

"Burn him! Burn him!" chant the crowd as the lights go down. The white-sheathed Glooms dance over the hills. There is the steady beat of congas as the Fire Dancer twirls and taunts and finally torches the great white puppet. He is still moaning as his copper-colored hair goes up in flames.

Traditionally in Santa Fe, once Zozobra is burned, people's luck begins to change. Bad things stop happening. Love and prosperity prevail. The predictors can't wait. "Next year we should think about stuffing Zozobra with our old bug reports," says Karen.

At the end of October, Doyne, Letty, their two sons, and daughter leave for a month-long trip through South America. The climax of their journey will be a solar eclipse viewed from the top of a fourteen-thousand-foot mountain near Chile's Atacama Desert. This is the kind of cosmic event scientists are good at predicting, and Doyne has been planning this trip for a long time. It is scheduled to coincide with Swiss Bank's renewal of Prediction Company's contract and another tenfold ramp in the size of the portfolio. Since Norman posted his chart describing the company's probability of ruin, they have regained all the money they lost in that first rocky week. No one is getting rich yet, but the models are performing as promised. Swiss Bank renews their contract for another three years.

The Holy Grail

Most so-called anomalies in the markets don't seem anomalous to me. They seem like nuggets from a gold mine, found by one of the thousands of miners all over the world.
— FISHER BLACK

When you have tremendous conviction on a trade, you have to go for the jugular. It takes courage to be a pig.
— STANLEY DRUCKENMILLER

As fall shades into winter, Prediction Company is cruising along, with the portfolio making a few hundred thousand dollars a month, when one day David Weinberger storms into the building and corners Norman in his office. "What the hell is going on?" he demands. "I just found out you guys cut the position."

Norman tries to calm him down, but Weinberger is on a rampage. The problem is Prediction Company's use of *interventions*. These occur when Prediction Company asks SBC's traders to override the models by cutting orders in half or abstaining from the markets altogether. This substitution of human judgment for computational analysis is supposed to be reserved for rare occasions, such as major "number days," when changes in interest rates are being announced. News events can drive the markets in unforeseen directions, and it sometimes seems prudent to stand clear of these oscillations.

"This is no time to lose your nerve!" Weinberger yells at Norman, who is running the company alone while Jim McGill is out of town and Doyne is traveling in South America. "This is an experiment.

You're building an automated system. You don't put your finger in the middle of it and diddle the results. You play it where it lays."

"We're also a business," Norman thinks. "I'm not keen on going broke, just to keep the experiment pure."

Norman knows that when Weinberger starts shouting, which seems to be happening with increasing frequency, the best response is to roll into a ball and sit there like a hedgehog. He will keep nipping at you, but sooner or later he goes away. Weinberger has other local diversions that keep him busy. He follows the local blues scene, and he sits on the board of the Santa Fe Institute, where he is advising them on how to build an artificially intelligent, computer-simulated stock market.

"The one thing we demand in our business is absolute discipline," shouts Weinberger, pacing from wall to wall, jangling the change in his pocket. "This is where the rubber hits the road. If you don't think your car is ready to race, you put more stuff under the hood. You don't take it off the road."

"Discipline," Norman agrees. "We could all use some discipline."

"Interventions might be appropriate for major number days. But in no way should this be a seat-of-the-pants decision."

"Dave, you're right," says Norman. "There's been a lack of process around here. I was at the helm. It's my fault."

Actually, it is everyone's fault. As they read the news feeds and follow the markets, they sometimes think their judgment is better than that of their models, and, oddly enough, the more money the models make, the more inclined people are to override them. This sometimes happens to gamblers who become increasingly timid with each winning bet.

A typical interoffice e-mail reads: "Tuesday through Thursday of this week are Japanese holidays. This means that the Japanese yen futures market may be thin, and when the Japanese come back into the market on Friday, there may be a big gap, comparable to a weekend gap in prices. So what do we do? We could do nothing, or we could close out our position on Thursday, in anticipation of a big gap. Comments?"

Responses are divided between people saying, "Close out the

position," and other people saying, "Don't intervene; Japanese holidays were never removed from the back tests, so we shouldn't start now."

Attempts to outguess the market are limited by the fact that no one really knows how markets adjust to news or holidays or economic events that often come from unforeseen directions. Even George Soros sometimes gets it wrong. In February 1994, misreading the political tea leaves, he lost six hundred million dollars with a bad bet on the Japanese yen. While Prediction Company is debating the effects of the upcoming Japanese holiday, it is not the holiday, but a surprise intervention by the U.S. government and fifteen allies that moves the market. They dump their yen holdings and send the Japanese currency into a free fall as part of a coordinated strategy to prop up the dollar. So, yes, the market gaps, but earlier and not in the direction and not for the reasons that anyone predicted.

After the yen, another e-mail exchange focused on the Deutsche mark, which was likely to be affected by a second attempt to prop up the dollar. Doyne and Norman placed a conference call to Weinberger.

"Sure, a G7 meeting next week means higher volatility," he said. "But I can't tell you which way it's going to go. I'm skeptical about intuitive hunches on direction, and I'm adamant about not wanting to lose pure model performance."

"Some of us are feeling that our research is lagging behind what we intuitively understand," Norman countered. "We're hanging out with big positions when major events are brewing. I think we have to work a lot harder on risk control."

Weinberger agreed. "You need the freedom to scale down from time to time."

Out of this conference call came guidelines for how Prediction Company, on rare occasions, would override its models. Now Weinberger is storming around Norman's office because of a misunderstanding about how to apply these procedures. He knew in advance that Prediction Company was cutting the portfolio's position in the oil market, but he thought the bet was being cut in half, while in fact it was being reduced by three-quarters.

If Weinberger is ticked off by this misunderstanding, imagine what he feels when he learns that Prediction Company, the following day, without being able to reach anyone in Chicago by phone and clear it in advance, unilaterally decides to crank down the T-bond model. Weinberger goes nonlinear. He storms back into the office. This time it is Doyne, newly returned from South America, who gets cornered for the tirade.

The upshot of Weinberger's second visit is that Prediction Company is supposed to work more closely with SBC's traders and keep their interventions to a minimum. This is another important step on their way to becoming seasoned professionals. Later, when they study the numbers, they realize that the portfolio would have made a lot more money if it had been allowed to run hands free.

"You have to be willing to put all your money in the pot and see what happens," Doyne says. "Our roulette experience is good training for this business. It's scarier than hell, hanging out in the breeze, but when the odds are in your favor, you have to swallow your fear and hold on."

In the three days before Thanksgiving 1994, the portfolio makes a few hundred thousand dollars in profit. By the end of the year it is solidly in the black. "It's almost scary how our predictions are coming in right on the money," William muses, standing at the trading console and watching the numbers tick down to the last trade of the year. He cracks a big smile as the models record their final win. To celebrate their good fortune, Doyne hosts a New Year's party at his house. With Stephen Pope on piano, Karen's husband, Jeff, on mandolin, Doyne on harmonica, and Chris Langton on guitar, everyone joins in to sing "Wabash Cannonball" and "Bye-Bye Love" as they work their way through a case of champagne.

The Swiss Bankers are returning to Santa Fe at the end of January 1995. To prepare for their arrival, there is a big push to get the old models functioning faultlessly and new models on line—including models for trading equities on the New York Stock Exchange. Everyone charges with renewed vigor into model building, debugging, coding, and other tweaks to the system. After a couple of weeks of working fourteen hours a day, it begins to dawn on them that there are not enough warm bodies to go around.

Jim Nusbaum is working hard to keep the prediction engine running in Santa Fe while moving duplicate technology to Chicago. Karen Lawrence and Doug Hahn are swamped with requests to build special features into the new 2.0 software release. This is the long-awaited research and production code that will replace the amateur programs hacked together in the company's early days. Laura and Jan are trying to balance the books. Sonia is buried under a mountain of dirty data. Jim McGill is flying around the world searching for numbers. Doyne, Norman, and William are burning the midnight oil trying to hack new models into production.

Prediction Company is beginning to resemble the old Eudaemonic commune at 707 Riverside Street during exam week. This is when studying for tests and writing research papers coincided with getting ready to hit the casinos in Las Vegas. The dining room table would be buried under a mess of computer chips and homemade mother boards. The air smelled of solder, and meals got reduced to grazing out of the refrigerator. At Prediction Company the computers are no longer homemade, but the kitchen is getting heavily grazed. It is stocked with the usual hacker brain food: bags of Famous Amos chocolate chip cookies, a whiteboard covered with equations, a large jar of organic peanut butter, three containers of Breyer's ice cream, four boxes of file folders, one pound of Sumatran coffee beans, a broken coffeemaker, four bottles of champagne, and cabinets filled to overflowing with granola bars, chips, crackers, and other munchables.

The day before the Swiss Bankers arrive, Prediction Company cranks into high gear. Researchers, looking like particles in Brownian motion, jog in and out of each other's offices, speaking in rapid-fire bursts about *fat tails*, *standard deviation*, and *degrees of freedom*. They draft research reports, fine-tune their models, sift through mountains of data, and put the finishing touches on overhead projections that show Prediction Company poised, as McGill jokingly says, "to conquer the world."

An hour into the morning the sweaters come off. By 10:30, Doyne,

Norman, and Jim McGill are headed over to the Aztec Street Café for their second or third hits of caffeine. The Aztec is filled with the usual nose-ringed travelers dressed in silver chains, baggy pants, and spiky, bleached-blond hair. The predictors retreat to the back room and shove together some tables for an impromptu director's meeting. They are trying to figure out the best strategy for getting SBC to open its purse. They want to move into stock trading on the New York and other exchanges, which requires an increase in their draw, the money they tap to cover their operating expenses in Santa Fe. They are looking for enough cash to throw seven extra bodies at the problem.

"Here are the most likely revenue numbers if we do another tenfold ramp," says McGill.

Norman studies McGill's projected P&L. "It looks like Swiss Bank stands to get a huge win."

"I have a question," Doyne asks. "Before we start talking about increasing our draw, do we give them the vision first and then hit them with the numbers the following day? Or do we give them the numbers straight off, so they can talk among themselves and figure out where they're going to get the money?"

"I recommend we give them the entire package all at once," McGill says.

He shuffles together his papers and rises from the table. "It's really pretty simple," he sums up, with an uncharacteristic grin on his face. "Basically, we're offering them the chance to make a lot more money with a lot less risk."

The secondhand clothing store that used to occupy the ground floor of Prediction Company's building has been replaced by the offices of a psychic called White Buffalo. The bearded man who runs this enterprise sells crystals and incense; he also offers something called White Buffalo Transforms, which take place in a small room in the back of his store. According to his literature on the subject, these transforms are capable of channeling "healing energy to your physical, emotional, and spiritual wounds."

Every so often one of White Buffalo's clients accidentally walks into Prediction Company. They barely get through the door before being shooed back downstairs. "We're developing strong psychic powers of our own," Karen jokes. "White Buffalo transforms. Prediction Company transforms. There's powerful stuff working here."

Norman, sitting in the V-shaped cockpit formed by the intersection of two computer screens, is almost lost to sight as he works at a desk piled high with books and papers. It looks as if every volume on his shelves, except for his 1911 Eleventh Edition Encyclopaedia Britannica has been summoned for consultation. Norman's leonine head of hair is shaggier than usual. His metal frame glasses are out of true, making him look like a cubist painting with each eye on a different plane. Also out of true are Norman's teeth, which got halfway through orthodontia before he ran out of money. The hardware stayed in his mouth all through college until one day Doyne picked up a pair of needle-nosed pliers and removed the metal himself.

Doyne is also looking a bit wooly. He is dressed in blue jeans and hiking boots and an oversized sweater fraying at the sleeves. His graying hair curls down the nape of his neck. He is spending too many hours staring at his computer screen, puzzling over model decay, while Clara snores at his feet.

Laura, taking a break from feeding the photocopy machine, where she is compiling a fat briefing book for the Swiss Bankers, collects everyone's lunch order. When the sandwiches arrive, the boardroom table hosts a free-floating seminar on nonlinear modeling techniques. Doyne, gesticulating with his hands, slices the air into seventy-dimensional state space. Then he unwraps his sandwich and dumps his bean sprouts on the table. "I can't stand them," he declares. "That was the worst part of living in California. They put sprouts on everything."

The predictors beaver away until nightfall, when Doyne and Norman go home for dinner. They read bedtime stories to their children and return to the office by 10:00 P.M. Meeting Norman on the stairs, Doyne drops into his office for a chat. "Weinberger doesn't seem too enthusiastic about increasing our draw."

"Swiss Bank got hit by the collapse of the Mexican peso," Norman suggests.

"I doubt they're hurting for a spare million," says Doyne. "But you never know."

Then Norman goes on. "Weinberger has been talking to me about what he calls the 'million monkey' phenomenon. How do they know we aren't making a million models and just showing them the best ones? Our profit was good for the year, but for all they know it could have been luck, rather than skill."

"This is why we need to raise our draw," says Doyne. "With more people we could develop better tests for knowing whether we have winning models or lucky monkeys."

Next day the New Mexico sky is a turquoise gem shining over white-shouldered mountains. A golden sun lights the Rio Grande valley, in which rise solitary plumes of smoke from the wood fires that are warming people's houses. People gather down at the Eldorado Hotel. Since the hotel is full of conventioneers, the predictors are scheduled to meet in the wine room, a small chamber off the bar, which is filled with racks of wine and champagne. "I guess they think we're going to be celebrating," jokes McGill.

Seated at the round table in the room are David Weinberger, Clay Struve, and Craig Heimark for Swiss Bank, and Doyne Farmer, Norman Packard, and Jim McGill for Prediction Company. Weinberger is wearing a new pair of cowboy boots, a silver belt buckle, and the leather vest of a riverboat gambler. Struve is sporting a red sweater, which matches the color of his face after a weekend skiing at Taos, sixty miles north of Santa Fe. Heimark is trying to look casual, but he can never quite pull it off. Is it the white socks and sneakers, the beeper attached to his waist, the wire-rim spectacles, or the brown hair worn schoolboy short that make him look like a cubicle guy in the Dilbert comic strip?

The first slide McGill projects on the screen is Prediction Company's 1994 financial statement. "Here's what we said we were going to do last year. We were late, but we got a lot of it done." He runs down the numbers, which include a tidy profit from running the system in "test mode." They took in another three hundred thousand dollars from the sale of stock to Weinberger, Struve, and Heimark, who today are wearing multiple hats as clients, stockholders, board members, and fellow researchers looking over the results of their more scientifically advanced colleagues.

Norman steps up to talk about their current projects. "We regard ourselves at a turning point," he declares. "I want to give you a vision of what's possible for the coming year." He projects another set of numbers showing the face value on the Prediction Company trading position, its leverage, and return on invested capital. The portfolio trades a few hundred contracts a day in markets ranging from oil futures and currencies to bonds and stock indexes. Even at this early stage, the positions can get hefty, and SBC is on its way to becoming a major player in Prediction Company's chosen markets.

Next, Doyne discusses where the company is headed. He flips through his overhead projections, one after another, driving through the data like a prosecutor going for conviction. His first slide compares the estimated and actual performance of their models. "We're dead-on with our estimates," he announces.

"I keep getting on my soapbox about diversification," he declares, pointing to a big stack of research reports and documents marked *Company Confidential*. "All the data agree. The key to success is to spread ourselves over as many models as possible."

"The trick is to build good, believable models," Weinberger interjects. He begins quizzing Doyne about Prediction Company's modeling "platform." This is the automated, all-purpose tool they are building to predict the markets, any market. The phlegmatic Struve looks on with the occasional nod, while Heimark keeps an eye cocked for discrepancies in the numbers. He peers at Doyne's charts through the bottom of his spectacles while leaning back in his chair with his hands behind his head.

Doyne projects a schematic diagram of the company's Deutsche mark model. "We create a little competition among the dozens of models that are working inside this one big model. We round up the usual suspects and let them trade against each other. It's survival of the fittest, with the best models reproducing themselves and populating each successive generation of models with more of their offspring."

Doyne moves to talk about Prediction Company's recent thrust into equities, or stock, trading. No one knows why the stock market rises and falls, but a common explanation goes like this. *Profit takers* drive down high flyers. *Value investors* prop up laggards. Standing in the middle, between the fickle and the faithful, are the *arbitrageurs*, who have no opinion on the subject, save for wanting to collect a toll each time market prices are realigned. Prediction Company hopes to become one of these toll takers.

"Here's the quick and dirty Prediction Company stock classification system, which was done in a couple of hours, without knowing diddly about stock trading," Doyne says of his preliminary results. The model is pure vanilla, and artificial vanilla at that, but it looks promising. Our idea is to build models for predicting a thousand of the most heavily traded stocks. The goal is to automate the modeling loop and develop universal modeling methods."

As one listens to Doyne talk about it, Prediction Company begins to sound like a creature endowed with a voracious appetite for numbers. It hungers after data to be analyzed, massaged, cleaned, transformed, patterned, mapped, propagated, networked, graphed, plotted, and scattered through state space until out the back of this massive number crunching machine comes another number indicating *buy* or *sell*. This number gets fed down a pipe into the trade engine, which is designed to function like a perpetual motion machine for spitting out predictions and accumulating a steady stream of money. Prediction Company hopes to buy two hundred thousand dollars worth of new data this year. McGill is advising SBC to buy another big chunk of numbers. He is riffling through every drawer in the bank, pulling out figures. He is scaring up historical data back to the Pharaohs. He is buying one of everything from all the world's real-time data providers, while trying to beam as much of this information as possible back to Santa Fe. The trick is to take this river of numbers and tease out of it the little eddies of turbulent fluid flow, stranger-than-strange attractors, fractals, or other patterns that look even momentarily significant.

The new technologies for playing the markets are data-driven. They soak up huge amounts of bandwidth and computer power. They throw a net of numbers around the globe and use these numbers to map market forces as they shift in real time. These maps show money quivering over fault lines in the Asian economy. They show money accumulating on top of America's already towering mountains of capital. The predictors hope to speed up this mapping process, and maybe, in the best of all possible worlds, speed it up so fast that it shows a few seconds before it actually happens where the financial fault lines are likely to open or the mountains begin to erode.

Every major player in the world's financial markets is developing these systems, if not out of conviction that they work, then out of fear that someone else will make them work. The competitors are locked in an arms race, with victory going to the technologically swift. Even the slightest mathematical advantage will be leveraged into billions of dollars in profit. This happened after the opening of the options markets, when the Black-Scholes pricing model allowed smart traders to play the markets to optimal advantage. No one wants to miss the next bright idea. It takes deep pockets to play this game, but the potential rewards are enormous.

The roll call of players developing black-box trading systems includes Fidelity Investments, Morgan Stanley, D. E. Shaw, and Goldman Sachs, which is reported to have fifty researchers working on the problem. The most tantalizing rumors concern Jim Simons, a former professor of mathematics at the State University of New York at Stony Brook. His company, Renaissance Technologies, with forty-five employees and five hundred million dollars under management, made a 1994 profit of two hundred and fifty million dollars. His secret? A portfolio of thirty-four models predicting futures and options prices, combined with an equity arbitrage system of pairs trading in the stock market. "This is our stalking horse," says Doyne. "It's the level of performance we should be aiming for."

Clay Struve rocks back and forth in his chair. He has been trying for years to design his own stock trading system, and he thinks Doyne is on to something. He growls out the occasional comment in a low voice that rumbles from deep in his chest.

"Watch out for the uptick rule," he warns. He is referring to the regulation that forbids Swiss Bank and other players in the markets from short selling a stock until there has been an uptick, or rise, in price. This prevents them from manipulating stock prices by piling on sell orders.

"What kills you in equities is leaks about takeovers," Struve adds. "This really knocks around the prices."

"Stock trading gives us millions of data points, with hundreds of degrees of freedom," Doyne continues. "We salivate over data like this because it allows us to go out and be scientists. This is one place where our analytic techniques are going to shine."

Doyne's next slide is a summary of Prediction Company's recent research accomplishments. These include releasing five predictive models, designing an automated trading system, and developing techniques that will soon put *thousands* of predictive models online. The final bullet on his slide announces that "The Holy Grail of market prediction has been reduced to a software problem." This elicits a few chuckles from the audience, and even Doyne admits that market prediction is not yet the kind of science that allows you to sleep peacefully at night. "I'm scared to death about hitting bad luck in live trading," he admits.

As Doyne charges into the final section of his presentation, he flips to a slide labeled "Long-Range Plan." "The idea is to increase volume and expand the domain of applications with a qualitative jump in technology. Instead of hiring fifty new researchers, which is what Goldman Sachs is doing, we have to automate."

Doyne holds up a copy of the "1995 Research and Development Plan" he has written with Norman and William. Known jokingly as the "Conquer the World" document, the plan includes some blue sky speculating about how Prediction Company, with "mature deployment" of its technology, could make SBC a nine-figure yearly trading profit. "The principal question facing us now is not, 'Can we do it?' " the report says, "but rather, 'Can we do it fast enough?' "

The document offers three scenarios for pumping more money into Prediction Company. "Plan A is a baseline scenario, in which we basically leave the relationship the way it is. Plan B calls for a modest increase in our draw, which will allow us to do some hiring. The company is stretched thin. We need more recruits in the ranks, or we face the danger of burnout among key employees. Plan C calls for a bigger increase in our resources, and hiring six new people during the coming year."

"I want to walk through the numbers tomorrow," says Heimark, fingering a copy of the plan. "This stuff looks encouraging, but we're not ready to finance a hundred-million-dollar company. You're talking about a strategic partnership that goes beyond our initial roles."

"I don't think we have a choice," Weinberger comments. "We have a barrier to get over. We have to spend the money to get a competitive advantage. Some smart guy off the street is not going to come up with the same edge for less investment."

"How big an operation do we want to build in Santa Fe?" Struve asks.

"We have to think big and aggressive," continues Weinberger, who is not interested in dwelling on the details. "This can have a major impact on the bank. But nothing proceeds until the models kick in and we start making money."

"That's why we're asking for more resources," Norman says.

Heimark counters with a warning. "There is no way the bank is going to make this kind of investment without demanding an equity stake in Prediction Company. As soon as my colleagues catch wind of these numbers, they're going to be all over you. They're going to want to own you."

The all-day meeting in the wine room is followed by dinner in a private dining room at the Eldorado. Doyne, still keyed up from describing how Prediction Company is going to conquer the world, launches into an equally impassioned defense of scientific research. He mentions three of the twentieth-century's great research laboratories: Bell Labs, which invented the transistor, Xerox's Palo Alto Research Center, which invented the computer mouse and "windows," and Los Alamos National Laboratory's Theoretical Division, which designed the atomic bomb, before moving on to complex systems research. "Could you build a Bell Labs, a Xerox PARC, or a T Division in the belly of one of today's big corporations?" he asks.

"No, you couldn't," everyone agrees.

"Today's corporations are bottom-line, P&L-driven. The only person who could do it is a Medici with an aesthetic appreciation for pure research. Maybe a Bill Gates," says Heimark.

"There is every reason in the world why Swiss Bank should fund a pure research effort," Norman counters. "This would give them a competitive edge over the half dozen other people with pockets deep enough to embark on this kind of program."

"You have to be careful when you make this kind of assumption," Heimark warns. "Bell Labs invented the transistor, but they didn't make any money out of it. The money was made by Sony. Xerox PARC developed "windows," but bad management never allowed them to capitalize on it. The idea didn't get off the ground until key people were hired away to work at Apple, Sun, and Microsoft."

"Between pure research and getting your ideas into production lies a demilitarized zone," says McGill. "Obviously, it's not easy finding good managers who are willing to work in this zone."

The debate has no motive other than livening up the dinner conversation, but at least one observer wonders if Prediction Company is testing the trustworthiness of its business partners. Is Swiss Bank going to nurture research or count beans? Are they patient or profit-driven? Who at this table is going to argue Prediction Company's case in the Basel boardroom?

At 9:30 the following morning the Swiss Bankers show up at Prediction Company to announce which of the three plans they are adopting. Are they tossing another million dollars into the pot, or leaving it to simmer on the back burner?

Snow is drifted under the pine trees from a midnight storm, and the air is filled with large white flakes of what look like confetti. But the storm is quickly blowing over. Great shafts of sunlight are beaming down on Santa Fe's chocolate buildings, and the clouds are giving way to patches of clear blue sky. The day is turning warm enough for the café crowd to lounge outside in their shirtsleeves until the high-desert sun goes down, and the temperature again falls below zero.

When everyone is seated around Prediction Company's boardroom table, David Weinberger opens the discussion. "No matter what numbers we come up with, there's always a chance we don't know what the hell we're doing. This may all be garbage. The problem is not amenable to pure statistical analysis. I'm trying to be realistic about this, but it will take years of practice before the intangible piece of the puzzle—how to deal with structure breaks and nonstationarity in the markets—goes away. This is never going to be just a numbers issue."

Doyne reminds him that if Prediction Company can hire more researchers, they should be able to resolve some of this statistical uncertainty. But he sounds more ambivalent than yesterday about asking for a million dollars. "Increasing our draw increases the hole we're standing in. My biggest fear is that we'll deliver the technology, but make a piddling amount of money after paying you guys back for your investment," he confesses. "Norman and I are not professional finance people. We want to get back to doing research. But the company is hung around our heads, and I question whether we'll ever be able to sell our stock. So I think you should know what I'm feeling today, that if I have to stick around Prediction Company another ten or fifteen years in order to get out my money, then I'll quit, even if I get nothing out of this."

"Welcome to the real world," Weinberger says. "This is what you go through to build a business. It takes a long time, and you don't necessarily get your money out right away. It's still early for you guys."

"Prediction Company can raise money one of two ways," advises Heimark, "either through equity or through debt." You want to do it through debt, that is, by borrowing money from Swiss Bank. But Swiss Bank may have other ideas."

The conversation turns to a discussion of warrants and other equity kickers that might mollify the bankers in Basel. "I've been coasting ever since we signed the O'Connor contract," chuckles McGill. "Now it looks like I'll be getting back to work."

"Prediction Company is still pretty fragile in terms of results," Heimark cautions. "We haven't quite got the science proved. This makes me wary about ramping the deal now. On the other hand, we have to do this quickly, or our competitive window is going to shut."

"What kind of numbers would make the bank happy?" asks McGill.

"We need a long-term structure defining the relationship between you and the bank."

"But the bank is unknown to us," counters McGill. "It's way down our list of options to think of getting into an equity deal with them."

"I think you're driven by fears," Heimark says. Then he acknowledges, "They may be reasonable."

"My biggest fear is that you guys say good-bye to Swiss Bank, and then I get a call from someone in Basel saying he wants to see me in his office Monday morning," McGill says. "Working for them half-time on the Data Project has allowed me to go out and survey the Swiss Bank terrain, and I can tell you, at the moment there's no coordinated data strategy across the bank."

"I'm ambivalent about which way to go," Doyne admits. "Last night I was pushing hard for Plan C. This morning I'm ready to throw up my hands and settle for the status quo."

"I thought we were buying Microsoft last night," Heimark jokes. "A big part of being a successful entrepreneur is to believe you're better than everyone else. So no one holds it against you, Doyne."

The meeting turns into a brainstorming session on how to get Prediction Company more money without Swiss Bank demanding to buy them.

"We're agreed. We're giving up our insistence on equity and going for a bigger draw," Weinberger concludes. "We want to fund you. It's progressing well. You're performing at the high end of our expectations. We know we're going to get you the money one way or another. So you should go ahead with the early stages of looking for more people to hire."

"Some of them may end up in Chicago," Heimark warns. "Ultimately this is where the production people should be located."

"Every time we get near Chicago, things take three times longer," McGill complains.

Heimark nods. "I know, and our costs in Chicago are double what you pay in Santa Fe."

"We have to start collaborating more," says Weinberger. He glances at his wristwatch with three dials. "We're also agreed it's time for you to ramp your positions again. In the past I've rooted for you guys to lose, so you'd learn that things can jump up and bite you in the face. Now I'm rooting for you to win."

Norman realizes he has just been given the green light to hire seven more people and make another tenfold increase in the size of the portfolio. "We need a game plan."

Weinberger rises from the table. "You figure out the best way to do it," he says. "But we're all agreed you *should* do it."

Three miles up the winding road that leads to the Santa Fe ski basin, at eight thousand feet above sea level in the piñon-covered foothills of the Sangre de Cristos, lies a Japanese bath house called Ten Thousand Waves. This mountain sanctuary consists of ten hot tubs, including one large, communal tub, which is shrouded in mist and blissfully calm; only the sound of prayer bells tinkling in the hills disturbs the silence. Early the following afternoon, when the Swiss Bankers have flown home, Doyne and Norman swap their clothes for kimonos and climb a shale footpath that winds among the hot tubs and cold plunges. When they reach the communal tub, which is empty at this time of day, they drop their kimonos and sink into the hot water. Steam rises around them. The water is fragrant with the smell of cedar. They drape their arms over the side of the tub and let their legs drift weightless in front of them. They stare into the mountains and try to empty their minds of everything except the feeling of having warm water lapping against their naked skin.

"Sometimes I think we're crazy to be doing this," Doyne says. "Every time we finish one of these meetings I have to fight an adrenaline letdown. Last night, I felt so lousy I was ready to quit."

"I hope we haven't alienated everyone," Norman says. "First we say, 'Put the pedal to the metal!' Then we say, 'Whoa, hold on! We don't really want to dig ourselves into a deeper hole.' "

Doyne leans back and stares at the sky. "Maybe we should just go with a modified plan A. Jim McGill's salary will be covered by the bank while he works on the Data Project, and we'll survive on the resources we already have. Every time we ramp our positions, we go down before we go up. It's already happened four times, and it will probably happen again the next time we push the button."

Norman nods. "We have to steel ourselves for a bumpy ride."

The valley below is darkening with purple shadows. The peaks above are lit with the red glow that made the Conquistadors named these mountains for the blood of Christ.

Doyne and Norman have been trying to make their fortune together for a long time. Their lives are like two chaotic systems, flung into wide-ranging orbits, scattered through numerous iterations, and then miraculously brought back into proximity with each other. The fact that they are in business again and still coaxing each other into wild schemes is compelling evidence that humans, like other nonlinear systems, are governed by sensitive dependence on initial conditions.

The last light is fading off the mountains, and the night air is turning cool. The stars come out and a crescent moon rises in the felt-blue sky. "I sometimes wonder if we took a wrong turn into an intellectual backwater," Doyne muses.

"What do you mean?"

"We've had to learn a lot of useless stuff, like the fact that a tick for a German-mark contract equals twelve dollars and fifty cents."

"There is a lot of arcane stuff in physics, too, like Reynolds numbers," Norman reminds him.

"But Reynolds numbers are universal," Doyne protests. "They know on Mars what Reynolds numbers are."

Norman laughs. "You're only saying this because you're a physicist."

"Getting a degree in physics is more mind-altering than three years of brainwashing by the KGB."

"One could say the same thing about founding a company," Norman replies.

Market Force

The stock market provides an excellent laboratory for testing theories.

—GEORGE SOROS

The truth is, succesful investing is a kind of alchemy.

—GEORGE SOROS

In the last week of February and the first week of March 1995 three unrelated events occur, with unforeseen consequences on Aztec Street. Prediction Company registers a fourfold increase in the size of the portfolio and logs its biggest profits to date. Nicholas Leeson bankrupts Barings, England's oldest bank, with a wild billion-dollar bet on the derivatives market. And Doyne Farmer meets George Soros at a Santa Fe Institute conference on feedback mechanisms in financial markets. "I very much enjoyed our meeting," Soros writes Doyne afterward. "With pros like you I had better retire."

During ramp week David Weinberger flies into town to wish everyone good luck. He strides through the door looking like a gunslinger off the plaza. He is kitted out in cowboy boots and a black leather vest with snaps made from old buffalo head nickels. He sports a trade bead necklace, two leather wrist bands fitted with amulets, and a silver belt buckle as big as a pizza tin. "You're on the radar now!" he shouts. "The risk committee is going to be looking over your shoulder, so this is a good time to start making some money."

No one knows whether it is Doyne's T-shirt or David's silver amulets or Prediction Company's synthetic neural networks, but ramp day is a winner and so, too, are most of the sunny days that follow. Prediction Company expects their review by the Swiss Bank risk committee to be a snap.

On February 27, 1995, Nicholas Leeson, a twenty-eight-year-old trader in Singapore, places a bet in the futures market. He is gambling that the Japanese stock market is going to rise, but because of the Kobe earthquake the stock market does not rise. It falls, just as it has been falling all week while a desperate Leeson tried doubling up his bets in a classic martingale. To cover his losses on the Nikkei 225 stock index, Leeson has been receiving huge amounts of money every day. "The senior people in London that were arranging these payments didn't understand the basic administration of futures and options," he claims. "They wanted to believe."

At the end of the week, when his bets are called, Leeson has lost one billion three hundred million dollars. This is more than Barings is worth, and the Queen's bank, a venerable two-hundred-thirty-two-year-old institution, closes its doors. One should note that George Soros, having wagered on the other side of Leeson's bet, finishes the week with several hundred million dollars extra in his pocket.

Leeson, the son of a plasterer, who fashions himself a kind of working-class hero, goes on the lam. He flies to the sultanate of Brunei and is trying to get back to England, whose jails he considers preferable to those in Singapore, when he is apprehended during a layover in Frankfurt. On being returned to Singapore, the unrepentant trader receives a six-and-a-half-year prison term and publishes a book, in which he claims that his Eton-educated bosses were more ignorant than he was about derivatives trading. "They never dared ask me any basic questions since they were afraid of looking stupid about not understanding futures and options."

Leeson's book is a parody of all the tomes that describe successful stock market strategies. In order to *lose* money in the markets, one should begin by thinking of oneself as James Bond playing with Nikkei 225 stock indexes instead of poker chips. "There are a lot of similarities with gamblers," Leeson says of derivatives trading. "Unfortunately, I lost more than I won. As stupid as this sounds, none of this is really real money."

Leeson's is one in a string of phenomenal losses registered by derivatives trading in the 1990s. The most shocking blowup for Americans is the announcement on December 5, 1994, that Orange County, California, is declaring bankruptcy after losing one billion seven hundred million dollars with a bad bet on interest rates. The largest municipal bankruptcy in history, the loss represents a thousand dollars for every man, woman, and child in the county. Robert Citron, treasurer of Orange County and the man responsible for this fiasco, claims in a hearing before the California Senate that he was an inexperienced investor who didn't understand what he was buying from Merrill Lynch when they sold him the interest rate derivatives that sank the county. Citron, who was also getting investment advice from psychics and astrologers, pretends to have been used as a "pigeon." "In retrospect," he babbles, "I wish I had more education and training in complex government securities."

As ludicrous as this sounds, coming from one of the nation's best-known municipal treasurers, the fact of the matter is that no one, not even the bankers who design them, know the value of certain derivatives contracts. Frank Partnoy, a former derivatives trader at Morgan Stanley, describes in his book, *F.I.A.S.C.O: Blood in the Water on Wall Street*, how his fellow salesmen used to joke about creating "weapons of mass destruction." After being suckered into buying these products, Morgan Stanley's clients would be "blown up" or have their "faces ripped off."

Partnoy describes the mad scramble at Morgan Stanley that followed the news that Orange County was declaring bankruptcy. The firm had sold them "structured notes," which are a complicated bet on interest rates. Morgan Stanley "didn't know how to value the structured notes precisely," Partnoy confesses. The firm mobilized dozens of traders, who worked through the day, trying to put a number on their liability. Their first guess was that Morgan Stanley would lose one billion six hundred million dollars. But by day's end, after flogging their Orange County notes on the open market, Morgan Stanley actually *made* money on the deal.

Derivatives are computer-age financial instruments that derive their value from another financial instrument. They come in a variety of flavors, including options, stock index futures, interest rate futures, and options on futures. Since derivatives derive their value from the underlying contract on which they ride, they are a bet on a bet, a financial freeloader whose primary economic utility is to make money out of money. Trading in stocks, currencies, and commodities is regulated by the federal government. Trading in the swaps and over-the-counter derivatives that ride on top of these markets is unregulated, out of sight, and off the books.

The tail attached to the dog is now ten times bigger than the dog itself. In 1995, when Orange County, Barings, Proctor & Gamble, Gibson Greeting Company, the Baptist Missionary Association of America, the Rock and Roll Hall of Fame, and municipalities from San Diego County to Saint Petersburg, Florida, were having their "faces ripped off," the total notional value of derivatives contracts traded in the world was fifty-five trillion dollars. This is greater than the total value of global trading in stocks and bonds, which in 1995 was thirty-five trillion dollars, and far greater than the annual gross national product of the United States, which was seven trillion dollars. The bulk of this derivatives trading—forty-seven trillion dollars—was done over-the-counter in interbank deals arranged by Nick Leeson with speed-dialers. Traders are playing with what economists call *hot money*: billions of dollars in investment capital that can be yanked in seconds out of Mexico or some other part of the world that yesterday was an "emerging market" and today is a financial dead zone. By 1999, the notional value of derivatives trading was up to ninety trillion dollars.

On April 12, 1995, Procter & Gamble, maker of Ivory soap and other wholesome household products, announces a loss of one hundred and fifty-seven million dollars on interest rate swaps. "Derivatives like these are dangerous, and we were badly burned," intones P&G's president, who says he plans to sue the bankers who put the dice in his hands.

The day after Procter & Gamble impugns derivatives, George Soros is summoned to Washington, D.C., to testify before the House Banking Committee. The committee is holding hearings on "Risks that Hedge Funds Pose to the Banking System." This narrowly focused topic disguises their real interest in learning why so many municipalities and company treasuries are getting blown up by derivatives.

The day of his testimony, Soros looks quite natty in a double-breasted pinstripe suit and paisley tie. He is an energetic, squarely built man with a furrowed brow, angular chin, and thin lips, around which hovers the hint of a knowing smile. His hair is cut *en brosse*. He wears tortoiseshell spectacles and speaks with the accented voice of a Hungarian refugee.

With great patience, Soros tutors the congressmen on hedge funds and modern finance. Hedge funds are investment pools that reward their managers on the basis of performance rather than with a fixed percentage of assets under management. Hedge funds, unlike other institutional investors, are free to sell short and play with highly leveraged options, which are two forms of betting with other people's money. Soros, for example, gears his investments at a ratio of five-to-one, which means he is betting five times more than the eleven billion dollars contained in his already ample fund. Hedge funds make dozens of simultaneous bets, long and short, in a balancing act that attempts to hedge out risks in one market with potential gains in another. Hence the name.

The other noteworthy feature of hedge funds is that many of them exist offshore. The term is not to be taken literally, as *offshore* includes parts of Luxembourg and London as well as the Bahamas and Curaçao, where Mr. Soros's Quantum Fund is officially located. *Offshore* means nothing more than operating free of taxes and domestic banking regulations. It is a peculiar feature of global capitalism— best understood, no doubt, by the rich people who take advantage of this oversight—that a sizable chunk of the world's investment capital circulates in the pirate economy of Captain Kidd. The Cayman Islands, for example, where Blackbeard is thought to have buried his treasure, are a mere speck of Caribbean sand inhabited by thirty thousand people. But the islands are home to more than five hundred licensed banks, and they rank as the fifth largest nation in the world in terms of booking bank loans. Of the seventeen trillion dollars in the global pool of investible wealth, roughly one-third of this amount, or six trillion dollars, is held "offshore," and one-third of *this* amount sits "offshore" in Switzerland in numbered accounts.

"I must state at the outset that I am in fundamental disagreement with the prevailing wisdom," Soros begins his testimony. "The generally accepted theory is that financial markets tend toward equilibrium and, on the whole, discount the future correctly. I operate using a different theory, according to which financial markets cannot possibly discount the future correctly because they do not merely discount the future; they help to shape it. In certain circumstances, financial markets can affect the so-called fundamentals which they are supposed to reflect. When that happens, markets enter a state of dynamic disequilibrium and behave quite differently from what would be considered normal by the theory of efficient markets." *Dynamic disequilibrium* is a polite way of describing panics, crashes, booms, busts, and other manias from which neither hedge funds nor congressmen are exempt.

Although he profits handsomely from instability in the financial markets, Soros is no wrecker. In fact, he disagrees with his laissez-faire colleagues who believe that market forces alone should rule the world. He knows how arbitrary, fickle, self-reinforcing, and self-destructive these forces may be, and he is continually calling for the creation of an international central bank, which will maintain stable currencies, reorganize debt, and guarantee an adequate flow of credit to the world economy. This proposal is far too radical for the House Banking Committee. So Soros limits his comments to topics they can understand.

"Frankly, I don't think hedge funds are a matter of concern to you or the regulators," he lectures his audience. But what *should* be a matter of concern to them is derivatives. He enumerates the "dangers" involved in "the explosive growth of derivative instruments." "There are so many of them, and some of them are so esoteric, that the risks involved may not be properly understood even by the most sophisticated of investors," he warns. "Some of these instruments appear to be specifically designed to enable institutional investors to take gambles which they would otherwise not be permitted to take. . . . One of the driving forces behind the development of derivatives was to escape regulations."

Trading in derivatives magnifies market volatility, says Soros, because it is basically a form of trend following—buying in response to higher prices, which pushes them higher, or selling in response to lower prices, which pushes them lower. Derivatives thereby reinforce whatever trend is dominant in the market. He himself uses them "sparingly," and his Quantum Fund is actually *beneficial* to the markets, he claims. "Our activities are trend bucking rather than trend following. We try to catch new trends early and in later stages we try to catch trend reversals. Therefore, we tend to stabilize rather than destabilize the market. We are not doing this as a public service," he adds. "It is our style of making money."

While believing that equilibrium is a myth, Soros disputes the other assumptions in classical economics as well. Perfect competition, perfect knowledge, and the laws of supply and demand are not laws, he says, but feedback mechanisms based on market influences and human perception. He published a book in 1987, *The Alchemy of Finance: Reading the Mind of the Market*, in which he presents his economic theory of "reflexivity." As fascinating as it is, the book is acknowledged by Soros to be a failure. Instead of presenting an economic theory, it hints at the first steps toward a theory. His work, not yet a science, has to content itself with being an *alchemy* of finance.

In 1994, several months before he meets Soros in Santa Fe, Doyne begins working on a scientific paper that he hopes will transform economic theory by proving that Soros's alchemical hunches are correct. The paper is originally intended for delivery at a conference on physics and finance held in May 1995 at the University of Chicago, but with the argument unfinished and Doyne suffering from the flu, he decides to skip the conference.

Called "Market Force, Ecology, and Evolution," the paper argues that financial markets are an emergent, self-organizing system whose behavior is closely related to that of biological systems. Markets comprise an ecology of traders who inhabit various niches. Trend followers interact with value investors, chartists with fundamentalists. Investors survive by following different strategies, and the interaction of these strategies is what creates patterns and oscillations in the markets.

The premise for Doyne's paper was inspired by talking to traders. He noticed there was one thing they all agreed on. Buying tends to drive the price up, and selling tends to drive it down. He realized this could be stated in mathematical terms and used as the basis for a theory of financial markets. The result is a weaker form of the usual laws of supply and demand. It is similar to Soros's principle of market reflexivity, but more quantitative and explicit in describing the feedback mechanisms between people and prices. People use strategies to buy and sell. When the price changes, this triggers more buying or selling, which in turn influences the price. All this depends on perception and human psychology.

From this point of view the market is the aggregate of its participants. Prices are both the end result and means of communication. Once the strategies used in the market are given, the price dynamics are automatic. Different strategies, such as trend following and value investing, generate characteristic behaviors in the price that may or may not reflect fundamental values. Oscillations between bull and bear markets are a natural consequence. Trading strategies are like the genes that distinguish biological species. Traders using some strategies profit; those using others lose and go out of business. As a result, the population of strategies evolves with time. The markets are an evolving, self-organized process that seems to work better than available alternatives, even if it is not perfectly efficient.

Doyne first learned of the efficient markets hypothesis from his brother-in-law Cookie Gibb, an investment banker working in New York, who has since retired to manage his own account. Back in the early 1980s, they were driving from Connecticut to Boston, chatting about their workday lives. Doyne swapped a thumbnail sketch of chaos theory for Gibb's description of modern economic theory, which struck Doyne as utterly fantastic. Surely, Gibb was talking about Erewhon not Earth. A theory that stated that markets are perfectly random, profits impossible, and people infinitely rational simply couldn't be true.

Doyne is not the only one to feel this way. The last decade has seen the blossoming of a new field called *behavioral economics*. It believes there are patterns in markets because traders behave in groups. They obey universal laws of human psychology. They share common elements in their backgrounds and biases; common information sources; and common emotions of fear and greed. They are not fully rational. They are overconfident. They have a poor intuitive grasp of statistics. They act in herds. They make decisions in a world endowed neither with perfect knowledge nor perfect competition.

One of Doyne's hopes for his new theory is that it will serve as a quantitative framework for behavioral economics. Because the theory is dynamic, without a priori assumptions that supply must equal demand, it allows for the investigation of how and when markets go to equilibrium and whether they are efficient. So far Doyne has found that markets are far from perfect. Financial evolution does not proceed to an optimum point and then just sit there. Traders exploit patterns in the market to make profits, causing old patterns to diminish and new patterns to form. As in biological systems, this results not in the attainment of some static endpoint, but in the evolution of successively richer and more complex strategies.

"My deepest interest in this is with the bigger questions about emergent self-organizing processes," he says. All this poking into the markets is really just a way to get at life's serious questions: questions about the nature of complex systems and the emergence of order in the universe."

Golden Handcuffs

I have watched the most able men and women of my generation, who might have created unexampled monuments in moral philosophy, mathematics or engineering, waste their lives in a prattle of non-accelerating inflation rates of unemployment or rather, since such matters cannot long occupy an educated mind, in interminable telephone conversations with their stockbrokers.

—JAMES BUCHAN

The business of financial management is selling black boxes and Prediction Company is blacker than most.

—EUGENE FAMA

Following their first flush of success in the early spring of 1995, Prediction Company hits a dry stretch of road full of potholes and dust devils. The oil model springs a leak; there is money flowing all over the floor of the World Trade Center. The 2.0 trade engine is still getting the bugs shaken out of it. The currency markets are getting whipsawed by the Fed. The risk committee is pestering them with questions, and then suddenly, out of the blue, comes more bad news from Basel.

During these difficult days on Aztec Street, rare good news comes in the form of two new employees, whose talents are divided between software and research. The new recruit in the software wars is Stephen Pope, the jazz pianist and former ski bum whose skills were employed to rebuild the Eudaemonic shoe. After his detour into programming toe-operated computers, Stephen helped manage the Advanced Computer Lab at Los Alamos. It was a glamorous job, jockeying the biggest computers in the world as they modeled climate change or atomic forces, but Stephen went through a "midlife crisis" after five years at the lab. "I wanted to tackle some *really* complex problems," he says. There was no shortage of problems at Prediction Company, and the place was a designer's dream; so he moved down to Aztec Street in February 1995.

Arriving at the same time as Stephen is the company's new researcher, Dave DeMers, a Falstaffian character with a walrus mustache, a thatch of gray hair parted in the middle, and the useful skills of a professional bridge player. Given to wearing wild, floral-print surfer jammies and T-shirts saying, "I am a professional. Do not try this at home," DeMers shares an office with William, where his round, cherubic face is the perfect antidote to William's Germanic angst. Born in 1954 into a family of French Canadian ancestry, DeMers, the son of an aerospace engineer, has a B.S. from Stanford in math. He has an M.B.A. from the University of California at Los Angeles. He has another degree from UCLA in law. And, finally, he has a Ph.D. in computer science from UC San Diego, where he used neural networks to control robot arms.

DeMers's first assignment at Prediction Company is to plug the oil leak. The prices at which their contracts are being traded diverge dramatically from the prices quoted over the satellite feed. The screen tells them the market is trading at price x, but as soon as they place their orders, the price gaps up to y or down to z. The larger their order, the more the price gaps. This gap, which traders call *slippage* or *market friction*, can be as much as two hundred dollars per contract. Multiply this by several hundred contracts per day, and one realizes how much money is draining into the NYMEX oil pit.

DeMers is dispatched to New York. In order to get into the pit and see what's going wrong, he is hired as a temporary employee by the trading firm that handles the bank's business. DeMers is watching the pandemonium, the usual hurly-burly of traders bellowing at one another and flipping hand signals off their noses, when the order based on Prediction Company's signal arrives in New York at 3:05 P.M. Amazingly, the market quiets. Everyone turns to look at the broker fielding the order. He is trying to buy three hundred futures contracts. Each contract equals a thousand barrels of oil. This three-hundred-thousand-barrel order, in a market whose total volume is about ninety million barrels, represents a significant piece of the business on the floor.

As soon as the trader makes a move, everyone piles onto his order, and the market price starts gapping up ten, twenty, thirty dollars, before the order is finally filled. Then, as soon as his business is out of the way, the locals let the price fall, having scalped out the profits that will allow them to go home happy men. "The guys were standing around licking their fingers," says DeMers. "By showing up at the same time every day, Prediction Company was just begging to get carved into lamb chops."

Doyne never intended to prove his theory of market force by dumping money on the floor of the NYMEX oil pit, but this scene provides a perfect example of how market price is pushed around by order flow. The cost of trading is defined as *market friction*. It has two components: the fees charged by brokers for their service and *market impact*, which, for big orders, is by far the larger of the two costs. Impact is caused by traders pushing the prices higher the more one buys and lower the more one sells. Doyne can demonstrate this phenomenon simply by graphing the size of the order against the price shift caused by this order. It is a neat little experiment, but an expensive lesson.

DeMers sees right away how to fix the problem. He recommends that a new broker be hired with a better position on the floor. If the broker could look from the futures pit into the neighboring options pit, this would allow him to detect the onset of order flow before it washed over him. Another solution is to break up orders and trade at different times of day. Soon the market friction generated by Prediction Company's signals drops nearly to zero, for savings of close to a million dollars a year. "The guys who used to stand around licking their chops don't know where the lamb has gone," says DeMers.

The summer of 1995 is unusually hot and dry. The mesas are scorched to dust, and even the pines flanking the Sangre de Cristos are beginning to burn as the forests around Santa Fe succumb to a rash of fires.

In August everyone at Prediction Company flunks their quarterly objectives. The software modeling group, still debugging a hundred thousand lines of code, misses the 2.0 release date; 2.0 is the cocktail of multivariate models that will execute Prediction Company's around-the-clock, around-the-world trading system. *Multivariate* means that it combines models from different markets. It also combines the company's historical and real-time data in one package, or *platform*, which is a standardized set of tools designed to crank out thousands of predictive models for everything from American stocks to Japanese bonds. The platform will have a suite of protocols for checking on holidays, half days, rollovers, and other glitches that get in the way of fault-free execution. It will have a trading calendar that updates itself automatically across the world's time zones. It will deliver real-time P&L to everyone's screens, while bundling back tests and live trading into one eye-popping package. "It's nice, very nice," says McGill, watching a demo. "But you get a zero for missing the release date."

The research group is faring no better. Their ambitious project to model the most actively traded stocks on the New York Stock Exchange remains a gleam in their eye while they wait for 2.0 to come online. The company is currently trading models in markets ranging from T-bonds and oil to Deutsche marks, yen, the S&P 500 index, and German bonds. The P&L is up a few million dollars, but every one of these dollars feels like it was laboriously hand-printed in Ye Olde Prediktion Shoppe.

The production people are also in the dumps. The plan to move Prediction Company's machinery to Chicago is getting nowhere. Jim Nusbaum, shuttling between Aztec and LaSalle streets, is in a black mood. "Swiss Bank has so many protocols and passwords it's slowing down the pace of development," he reports. "I recommend we nix the idea of moving production to Chicago and just do it ourselves."

Nusbaum also reports that Swiss Bank is in turmoil after their latest corporate acquisition. They bought O'Connor & Associates in 1991. Three years later they bought Brinson Partners, a Chicago fund management company run by former professor of finance Gary Brinson. Less than a year later, the restless Marcel Ospel, who engineered the Chicago purchases, is out shopping again. In the spring of 1995, Swiss Bank Corporation startles the financial world by announcing that it is buying S. G. Warburg, Britain's biggest investment bank, for one billion three hundred million dollars.

Warburg was a scrappy firm that outsmarted and eventually seized a lot of business from the bowler-hat set that used to run Britain's stodgier merchant banks. But now Warburg is falling to an even scrappier player: the newly Americanized Swiss Bank Corporation. In one fell swoop, SBC is becoming a leader in British corporate finance and one of the largest securities firms in the City of London. It is joining the "masters of the universe," the top-ten world players in investment banking—called merchant banking in England—which includes foreign exchange trading, underwriting new stock and bond issues, and piloting companies through mergers and acquisitions.

While Swiss Bank suffers from indigestion, Prediction Company develops its own ulcers in August. August is a good time for the Federal Reserve Board to intervene in the currency markets. Traders are on vacation, and the markets are thin, so the government's money stretches further as the Fed tries to prop up the U.S. dollar by selling Japanese yen and German marks. Central bank interventions are like the hand of God reaching into the Newtonian universe to jimmy the mechanism. This month, when the Fed reaches into the market and changes the settings, Prediction Company's models are on the wrong side, losing a bundle of money in the process.

Central bank interventions in the markets represent what economists call *external shocks*. No matter how prescient Prediction Company's models may be, they will never succeed in second-guessing the Fed. Prediction Company's only hope is that these shocks are relatively infrequent and that the company might, in time, learn to stand clear of them. In the meantime, William is waking up every morning worried that the Fed, the G7, the Bundesbank, or someone else with deep pockets is going to hammer the markets. He understands how the markets work. He is getting good at predicting closing prices and intraday market moves down to the minute, but he has no defenses against a monster who can reach into the mechanism and jimmy it at will.

What might be called the Leeson effect—the fear by senior bank managers that a rogue trader in some obscure part of the world will bankrupt them overnight—does not go unnoticed in Basel. The Swiss Bank Risk Committee starts paying more attention to the SBC Prediction Company portfolio. They like what they see. Prediction Company has a handle on risk. But the committee's increased attention to a previously obscure part of the bank provokes a chain of unforeseen events, as Prediction Company soon learns. On September 5, 1995, an SBC executive arrives in Santa Fe bearing a message from Marcel Ospel. Swiss Bank wants to buy Prediction Company. What do they think of the idea?

The bearer of the message is Joe Doherty, a former partner at O'Connor & Associates who has been transplanted to Basel. Of medium build, slow-speaking and thoughtful, without an ounce of bluster about him, Doherty projects the air of a family lawyer counseling clients on estate planning. He is a devoted family man in his mid-fifties, a former Chicagoan who now enjoys good French food and wine. He is casually dressed in loafers and sports clothes and appears to be doing nothing more significant than paying a social call on new friends in Santa Fe.

Doherty invites Norman and Doyne to dinner at Pasqual's, a small restaurant off the plaza. It has a communal table in the middle and smaller tables off to the side. The uncomfortable cane-backed chairs are soon forgotten once the barbecued duck and grilled salmon arrive. David Weinberger has flown out from Chicago to join them. Jim McGill is off on his yearly vacation in Hawaii. Doherty orders a nice bottle of wine and sits back to hear what Prediction Company thinks of Ospel's proposal.

"Our idea of what we're worth is likely to be a lot more than what the bank is willing to pay," Doyne begins. "And it's not until we've started printing large sums of money for the bank that we'll be able to convince the skeptics."

Norman mentions that what they're doing as an independent company may not be possible if they became part of a large investment bank. "You don't want to kill the goose that's just about to lay the golden eggs. Until we actually lay enough of these eggs that no one can say it was an accident, wouldn't it make more sense just to extend the existing contract?"

"Prediction Company is on the verge of cracking the problem and making some serious money," Doyne continues, tucking into his dinner. "Why don't we just wait and see how things play out?"

Then again, Norman and Doyne are willing to sell out if the price is right. The issue is hotly debated back at Prediction Company, while Doherty returns to Switzerland. "This is a preemptive strike," Norman speculates. "Our current contract ends on September 30, 1997, and they probably want to merge us into SBC before we have a chance to link up with someone else or go off on our own."

His face opens into a cockeyed grin. "You have to admit," he says, "there is some appeal to getting rid of downside risk. Becoming part of Swiss Bank could significantly increase our resources."

They begin trying to anticipate McGill's response. The considered opinion is that he will want to rush off to Switzerland and clinch the deal. "That's the kind of race this horse was bred to run," Doyne says, steeling himself for a confrontation.

As soon as McGill arrives back in Santa Fe, looking tanned and fit after two weeks of body surfing off Maui, he joins Doyne and Norman in the back room at the Aztec Street Café. The tonsured fringe of hair over his ears is bleached white from the sun, and he is sporting a new pair of black, wire-rimmed spectacles.

"Right now we have 'mind share' on Marcel Ospel's desk," McGill begins, sitting down with his espresso. "It may not be there two years from now. Every project has its moment in time."

Doyne sips a coffee-flavored Italian soda. "That's true. But two years from now the P&L could be off the charts. Rather than selling ourselves short as some unproved entity, we could be in a position to ask for a lot more money."

"It takes profits of fifty million dollars to get their attention," McGill says. "The big players bring in a lot more than that."

Norman fingers his coffee cup. "There's always the chance that two years from now we'll be doing *worse* than we're doing now."

"I just don't think it's in our best interest to become part of Swiss Bank right now," Doyne says.

McGill shakes his head. "If we go looking for another partner, we're talking about fifty qualified organizations. Only ten will be suitable, and five will have already been checked off the list. This will be a long, slow dance."

He keeps searching for the middle ground. "There are ways to structure it so you guys won't have to be here three years from now. You could hire four more researchers and turn them over to SBC as a functioning unit. I see a buyout paying ten million dollars to each of you and leaving you free to go off and do whatever you want."

Is the apple being held out to them really this tempting? Norman and Doyne do a quick calculation. "I don't think they're going to step up to this kind of number," says Norman, leaning back from the table.

Doyne agrees. "They're just not ready to pay this much at this point, and why should they? We're still unproven, and I don't believe they'd let us go without an earn-out clause, which would keep us hanging around for a good many years."

McGill sighs, stares off for a moment, then comes back on another tack. "If we don't go along with the buyout, SBC could build up a shadow organization. Two years from now, when the contract ends, they take our technology and start trading without us."

"I don't see this happening," Doyne says, shaking his head. "Look at how difficult it's been to move our operations to Chicago."

McGill gulps down the last of his coffee. "We're four years into this project, and the technology is still unproved." He is beginning to get testy. "As a start-up, we don't look like Apple Computer or Compaq. They had to rewrite their business plan every month in order to build new factories. Our 'product' isn't flying out the door. We're not where we want to be. We're behind the curve. In order to get the kind of return we're looking for, we need a breakthrough. We have a lot of tantalizing results and great projects in development, but we haven't hit a home run."

Outside the window, it is another sun-soaked morning in the desert. The day is heating up into the nineties, but with any luck, the thunderheads beginning to pile up on the mountains will bring rain by evening. Marcel Ospel, if he knew, would be surprised by how vociferously a few comments uttered in Basel are reverberating halfway around the world. McGill steps over to the counter to order another espresso.

Ever since Ospel's messenger arrived in town, Norman has been fantasizing about what he might do with some money in his pocket. He could pay off the unpaid balance on his credit cards or buy an apartment in Milan or some land in New Mexico, maybe an "inholding" in the middle of the Santa Fe National Forest.

The two partners stare at each other silently as McGill returns to the table.

"Maybe we should ask Doherty for an opening volley," Doyne says, as McGill sits down. "Although once we ask him for some figures, everything we do after that will look like a negotiating ploy. He might get pissed off if we take up his time negotiating and then in the end say, 'No way.'"

"If we force the issue of remaining independent we risk getting sandbagged by the bank," says Norman. "They could, in principle, avoid cooperating with us on anything."

Doyne turns to him and says, "We could always go back to the idea of becoming an investment fund. We could hang out our shingle and raise the money ourselves."

McGill shakes his head. "In that case you'll need someone else to run the company. I'm not interested in spending my life doing road shows for fund managers."

"We could raise a hundred million dollars," Doyne muses. "Make some use of the undue publicity we've gotten."

McGill isn't buying it. "Our personal utility functions are very different. It's a risk ratio question, but you're plugging different numbers into the equation."

"In the end this may come down to a philosophical argument, rather than a discussion of facts and figures," Doyne admits. "What I fear most is selling this idea short. We make a little money and quit before the technology takes off. Then I'll have to explain to myself how I *really* could have done it if I hadn't walked away from the idea too early. I'm afraid of repeating our experience with roulette."

"How about a partial buyout," Norman suggests, "where the bank becomes one among a number of shareholders?"

McGill shrugs. "That's like being a little bit pregnant," he says despondently. It's obvious now, this is one race this horse is not going to run.

After lunch, Doyne and Norman quit work early to hike up Atalaya, one of the mountains that ring Santa Fe. They pause on the ascent to look back over town. Far to the west, where the Jemez Mountains rise above the Rio Grande, Atomic City, as Los Alamos is called, can be seen clinging to the side of an ancient volcano, which used to be higher than Mount Everest, until it blew its top two million years ago, leaving behind a great circular caldera called Valle Grande. Some nice hot springs are tucked in the flanks of this old volcano, and it is one of Norman's favorite spots for mushroom hunting.

A laudatory article about Prediction Company ran in today's *New York Times*, and Doyne and Norman have been busy fielding phone calls from Bloomberg, *Wired*, and other reporters who want to do follow-up interviews. Fending off depression, Doyne says, "With press like this, we could raise our own investment fund."

"But gearing up to manage it is a pain," says Norman. "The regulatory process is a pain. Hiring a new president is a pain. I don't want to do fund management."

They trudge up the mountain, mulling over their options, computing the odds on their being bought out or starved out. Maybe Ospel's "mind share" will wander somewhere else. Maybe they can sidestep this momentary interest and drift back into being Swiss Bank's black box in the desert. They decide to adopt a wait-and-see attitude. McGill will be restrained from flying to Switzerland, but Doyne and Norman might pay a visit to Swiss Bank's London office later in the fall.

"Maybe they'll drop the buyout if we offer to extend our contract," Norman suggests.

Doyne agrees. "We'd like to say 'no' to a buyout now, but it's something we'd consider after a two-year contract renewal. This gives us some wiggle-room."

Norman and Doyne, still trudging toward the summit, look a bit cheerier now that they have thought of some way to appease the gods.

Rapid Divergence of Nearby Trajectories

There is a tendency to advance in complexity of organization.

—CHARLES DARWIN

A social space where money does not rule—we have never needed it more.

—PETER DRUCKER

In September 1995, Shelby Rose, an attractive young woman wearing a black leather dress, is sitting in Rafael deNoyo's lap, while holding a big cheroot up to his seamless, baby-faced mug. He takes a puff, exhales, and returns to telling her about financial forecasting and other lucrative aspects of global banking. They are sharing a bottle of red wine in a biker's strip bar in Salem, Oregon. DeNoyo is a regular customer, a big tipper, and friend of the family, so to speak. In fact, he is designing an erotic web site for Shelby, which will be hosted on deNoyo's own site for Dynamix Trading (www.DynamixTrading. com/shelby). Dynamix is the latest incarnation of deNoyo's financial forecasting business. To this mother site he is adding a raft of off-shore services geared toward "knowledge workers"—computer-literate investors who appreciate the benefits of one-stop, tax-free shopping for all their financial, gambling, and other needs.

DeNoyo had worked for the born-again Christians in Chicago as long as he could take it, and then he decided to strike out on his own. This provoked Joe Ritchie into suing him for breach of contract.

DeNoyo skedaddled offshore for one of his periodic sabbaticals in the Caribbean. He went to work for a Bahamian hedge fund run by Ed Bosarge. Bosarge, at one time, was also suing deNoyo, but the two former business partners decided to kiss and make up.

Also working for Bosarge in the Bahamas was Tom Meyer, the lanky jock who used to break the office furniture on Griffin Street. After leaving Prediction Company, Tom had gone to South America to play volleyball. Then he moved to Las Vegas, where he deployed Prophet, Prediction Company's genetic algorithm, to forecast the odds in football and basketball games. Tom's sports betting operation was turning a handsome profit until, for some reason, he, too, felt compelled to take the air offshore.

Down in Nassau, they specialized in de-encrypting data from satellite feeds, which allowed them to run their hedge fund on free numbers. One day Rafael decided to engage Tom, a born-again Christian, in a theological discussion. "He knew nothing about Christian apologetics," deNoyo concluded. Tom locked himself in his office and refused to talk to anyone in the company for a week. "He was such a prickly personality that it was hard to run a business around him," says deNoyo.

Late in 1994, when Joe Ritchie's ire began to cool, deNoyo slipped back into the States, to Salem, Oregon, his wife's hometown. He bought a wood frame house in a residential neighborhood and turned the garage into a satellite node flooded with market data. The space was divided into cubicles separated by bookshelves holding works on everything from anthropology and philosophy to parapsychology and theosophy. Into this "feed guru's" haven deNoyo installed colleagues who include a British video game programmer named Man Dick; an army captain who once worked as brand manager for Rely tampons; a former combatant with the Nicaraguan contras; an habitué of London fetish parties, which feature live whippings on the dance floor; and the former Miss Florida, who used to make infomercials such as "Unleash the Madman Within You," which Colin Ferguson watched over and over again before he opened fire in 1993 on a train full of New York commuters, killing six.

DeNoyo is trading a couple of million dollars for an offshore fund based in the British Virgin Islands. He is doing football picks. He is launching a financial data service. He is designing an online gambling site, and he is planning to open a bank in the Cook Islands. "I'm starting a new company in Las Vegas," he tells Doyne, when their paths cross again. They are down on the strip, sitting in deNoyo's favorite bar, but it is a quiet night, with deNoyo puffing his own cigar. "The company is going to provide real-time market data over the Net. Bandwidth is getting big enough and compression technology is getting good enough that I can pump out all the numbers as they arrive. The site is going to provide quantitative analytics and everything else you need to massage the data. I'm also starting a web-based casino company running out of the Cook Islands."

"Where are the Cook Islands?"

"To the left of Tahiti, off New Zealand. They have bank secrecy laws and all the stuff you need for offshore trading."

"I've never understood why anyone would bet in a casino, much less on the web."

"Yeah, I know," deNoyo agrees. "Even my eight-year-old daughter can see the flaw in the system. 'Daddy, when you play poker on the web, you can't see them deal the cards.' She's right. The 'dealer,' in this case, a computer program, can deal any cards it wants. Personally, I think you have to have a screw lose to gamble on the web. But, heh, we're providing a service people want.

"I'm also keeping a race book and doing sports betting over the Net. Casinos, brokerages, and banks all have the same back office," says deNoyo, taking a swig of wine. "My grand plan is to combine them all. You do your gambling and banking from your brokerage site, which is set up offshore in a numbered, anonymous account. Your money is always in play, gambling, investing, betting, whatever you want. We're running the games, supplying the numbers, keeping the books, filling the orders. You can make a fortune just by standing in the middle of the action." Later in the evening, after the floor show, deNoyo allows himself to wax philosophical. "More and more knowledge workers are going to decide to opt out of the tax structure," he says. "The economic cream is going to skim itself off the high-tax economies and move offshore. Once the Iridium satellites are up, bandwidth will be so cheap that you can telecommute to your job from anywhere in the world. So why live in hell, like New York City, when you can live in paradise, like rural Oregon? Sell hell, and buy paradise! This could change the nature of civilization," he concludes, with a last gulp of wine. "At least that's the plan, anyway."

As they are getting ready to leave and deNoyo is reaching into his pocket for a big tip, he offers a final thought on what the future holds for people like him and Doyne. "All the superwealthy people I've known have been assholes," he cautions, shaking his big, curly-haired head. "Great wealth leads to disassociation from reality. No one tells you the truth. Everyone kisses up to you, lies and strokes you, until finally you end up lost in your own world."

"Are you in immediate peril?" Doyne asks.

"I have too many plans," deNoyo laughs. "I'll always be running a negative balance."

At Halloween Jim McGill announces he is quitting. For someone used to wheeling and dealing in Silicon Valley, he has been under-employed in Santa Fe for a long time. There is no buyout to negotiate, and in September 1995, when Prediction Company celebrated its fourth birthday, his stock options became fully vested. He is being courted by two would-be employers. Swiss Bank wants to hire him full-time to become the company's data czar. He would move from consulting on the Data Project to implementing it. Morgan Stanley, the New York investment bank, is competing to hire him as head of engineering. He would design their systems for capturing financial data and delivering them to traders' desks. His salary and other perks will likely be over seven figures. Once the bidding war ends, McGill has to decide whether he is moving to New York or Basel.

At the beginning of December 1995 Doyne and Norman fly to London to meet the Swiss Bankers who are deciding whether Prediction Company is to be bought out, starved out, or left to its own devices. "It's a good idea to introduce ourselves," says Doyne. "Just to let them know we aren't some wild-eyed lunatics out in the desert."

On an unusually cold day, with a blast of arctic air plunging London into a deep freeze, they show up at Swiss Bank House on High

Timber Street. They pause on the banks of the Thames. They are standing at the edge of The City, the square-mile financial district that lies between the Tower of London to the east and St. Paul's Cathedral to the west. Above them, up a flight of steps, rises Sir Christopher Wren's Renaissance dome. Below them, on the embankment, huddle the smokers who have been driven from their buildings. Swiss Bank House is a modern structure, with a glass atrium looking out on the river. Doyne and Norman are given a tour of the trading floor and the neighboring financial exchanges, but the key event of the day is their meeting with David Solo and Andrew Siciliano.

Solo is the thirty-year-old wunderkind from Chicago who is in charge of *rates*, which is short for *interest rates* and includes everything from Treasury securities to interest rate swaps and government bonds. Some of these contracts are exchange traded, but the big money long ago left the official exchanges and moved onto the electronically interconnected trading floors of the world's dozen largest banks. Solo is a slim, dark-haired man, so hyperactive, so speedy and insistent in quizzing people on their numbers, that he "acts like he's on an IV drip of Dexedrine," says an acquaintance.

"There's no way we can do a buyout deal because there's no way we can agree on a price," Solo declares. "Prediction Company is still an unknown quantity. So instead of buying you, I propose we extend your contract for three years."

Doyne tries to hide his pleasure at remaining independent. "That's sort of what we had in mind," he says. "Extending the contract sounds like a good idea, except we were thinking of *two* years."

Siciliano, another dark-haired Chicagoan, who is the boss of SBC's foreign exchange trading, sits at the table with a sober expression. Every few minutes a senior manager slips into the room to give him some news or tell him about a major change in their position. "I don't like the fact that you're out in the middle of nowhere," he remarks. Norman explains that one of the reasons the company is successful in attracting and keeping good employees is precisely *because* it is out in the middle of nowhere.

A general discussion of how the predictors might become "more intellectually involved with the bank" turns into a review of the portfolio and its daily standard deviation. This is the sum of money that the portfolio can be expected to win or lose in the course of an average day's trading. It is already large and about to get larger.

Now that it looks as if the partnership will be extended, Solo and Siciliano allow themselves to get excited about what Prediction Company can do for Swiss Bank. Siciliano believes that automated market making is the wave of the future and that potentially this could be the most lucrative application for Prediction Company's technology. Doyne and Norman nod their heads in agreement. They already know that someday their system could be converted from predicting markets to making them.

Successful market makers get in the middle of the action and control the order flow in capital markets by keeping their spreads tighter than the competition. Market making is big business for Swiss Bank. They have hundreds of traders shuffling billions of dollars a year in spreads. This operation is run the old-fashioned way—by the seat of their pants. Market making for Swiss Bank and their competitors is a labor-intensive enterprise based on guesswork. But if the system could be automated and the spreads narrowed, the potential advantages are huge. Siciliano and other bank officials are of the opinion that automated market making could revolutionize finance.

Doyne and Norman field questions on other potential initiatives. The financial markets are in the middle of an arms race in developing new technology, and Prediction Company might hold one of the keys to Swiss Bank's success. They could automate Swiss Bank's execution services. They could become an historical and real-time data provider. They could design and sell software, become business consultants, manage a fund. With so many ideas popping around them, it is hard for Doyne and Norman to believe that four years ago they were sitting on folding lawn chairs in an adobe house in Santa Fe trying to learn where the Euromarkets are located.

They decide not to replace Jim McGill when he leaves. Instead, Doyne and Norman will run the company as copresidents. Jointly, they will manage research and model building. Doyne will handle operations. Norman will manage their business affairs. His first job on getting back to Santa Fe is to drag out the original contract and start discussing the points that have to be negotiated before Swiss Bank and Prediction Company can extend their partnership to December 31, 1999. Norman attaches himself to Joe Doherty in Switzerland through the old-fashioned umbilical cord of a fax machine, and they start the meticulous task of hammering out a new contract. On the day before Christmas 1995, a Saturday, he and Doherty seal the deal with a verbal handshake and congratulations for getting the job done so amicably. The terms are similar to what was discussed in London, and Prediction Company now gets a big financial boost for signing on to the end of the millennium.

Everyone is impressed by Norman's talents as a negotiator. If their first contract with O'Connor & Associates was a sweet deal, the new one is even sweeter. They get enough resources to hire seven new people and quadruple their computing power. The contract also calls for SBC to hire seven new employees in Chicago, who will be running Prediction Company's models. In the meantime, Norman goes shopping for a dozen Sun UltraSPARCS, their top-of-the-line machine, and posts e-mail announcements looking for researchers to hire. "We have no problem with our cash flow at the moment," he acknowledges, with a grin on his face.

Among the new recruits is Martin Casdagli, a British mathematician who once worked as Doyne's postdoc at the Santa Fe Institute and who has been consulting at Prediction Company for the last two years. Casdagli was recommended to Doyne as the "best British dynamical systems person in his generation." There are no academic jobs in England for mathematicians, not even the best of their generation, and not many in America, either; so when his postdoc ended, Casdagli went to Wall Street. He took a job at Tech Partners, the firm that had also tried to hire Norman. Tech Partners had some bright ideas about statistical arbitrage, but when it came to directional forecasting, they had only one winning card up their sleeve, a neural net model for predicting the S&P stock index. The model worked for a while and then it stopped working, until the company folded in 1992.

Norman and Doyne's copresidency quickly falls into a daily routine that begins with their meeting every morning at the Aztec Street

Café, where company business is dispatched over croissants, cappuccino for Norman, and six-herb tea for Doyne. Both Zeus Pelkey and David Weinberger admit to being surprised by how easily Norman and Doyne are running the company. They wondered if Norman, the abstract thinker, would be able to negotiate a deal. Would he care enough about the nitpicking details to write corporate reports? Would he crank out the paperwork and meet the filing deadlines? The new contract proves Norman's skills as a negotiator. His quarterly reports, corporate accounts, tax filings, spreadsheets, and other paperwork are as crisp and timely as anything produced by McGill. He is quick at analyzing options and weighing solutions. In fact, Norman is a polymath who conducts business as easily as he plays the piano.

Pelkey and Weinberger are equally admiring of Doyne's skills at running the company's operations. They are reorganized to function around projects that mix researchers and software developers working together seamlessly. Doyne takes over as head of the stock modeling group, which includes Martin, Karen, Stephen, Doug, and Laurens Leerink, a new researcher from South Africa via Holland and Australia. To everyone's surprise, the group starts exceeding their quarterly objectives, something that has never happened before. Meetings are limited to Mondays and forbidden from running for more than one hour. Friday afternoons are reserved for a "tea" that doubles as a research seminar. The trade engine is humming along. The model building process is getting automated. The new 2.0 software is solid. The production systems are finally moved to Chicago, and the stock trading system is about to kick into gear.

On Norman's forty-second birthday, March 26, 1996, the newspapers are filled with obituaries announcing the death of his cousin David Packard, who founded the electronics firm Hewlett-Packard. Norman is descended from Zachias Packard, who came over on the Mayflower. He is related to the car-making Packards and the computer-making Packards via the family homestead in Colorado. Norman's side of the family migrated southward from Colorado to Silver City, where his father taught junior high school mathematics. The other side of the family migrated west to Silicon Valley, where they made America's first great personal fortune from computers.

Norman had visited his cousin a few times at his various homes and ranches in California. He had not given much thought to David Packard, except as a distant family member, until he began running a company of his own. Admittedly, a twenty-person trading company in New Mexico bears little resemblance to an industrial powerhouse with thirty billion dollars in yearly sales. But Prediction Company has to be managed somehow, and David Packard's techniques seem germane. As described in his autobiography, *The HP Way: How Bill Hewlett and I Built Our Company*, Packard attributed their success to two deceptively simple ideas, which he called *management by objectives* and *management by walking around*.

Management by objectives is the antithesis of management by control. It is not a military, top-down approach, where people are assigned jobs to do. It is a system in which corporate objectives are agreed upon by everyone involved, both managers and workers. How these objectives are accomplished is delegated to the people responsible for meeting them. "It is the philosophy of decentralization in management and the very essence of free enterprise," wrote Packard.

Jim McGill had already tried to import this idea into Prediction Company. But McGill, the son of a career colonel who flew B-52s in the Strategic Air Command, was always a big believer in maintaining a proper chain of command. He ignored Packard's other idea of management by walking around. Packard had discovered the importance of getting involved in the production process when he was a young scientist working in the vacuum tube department at General Electric. The company had been throwing away large numbers of faulty tubes, until one day Packard rolled up his sleeves and joined the workers on the shop floor. Between them, they figured out where the company's written instructions were inadequate; other improvements resulted in GE's being able to make fault-free vacuum tubes. By the time Hewlett-Packard was a billion-dollar operation with manufacturing plants scattered around the world, Packard could no longer manage it by walking around, but he insisted up to the end that the company's executives work in doorless, open-plan offices.

During one of their conversations about how to run a company, Jim

Pelkey counsels Norman, " 'Management by walking around' doesn't mean you're a friendly guy who pokes your nose in everyone's business. It works best when you perform a peculiar inversion. You role play working for your employees. You sit down and ask them, 'What can I do to make your life more productive?' They may want you to run interference with Swiss Bank or buy them new tools, but employees get juiced on this kind of attention from management. That's your job, to get everyone charging forward together."

Norman is beginning to appreciate other similarities between his own life and that of his cousin. For years he had thought of himself as an abstract thinker, interested only in pure research. But now he is developing a taste for *praxis*: ideas in action. "The thought of using a spreadsheet was repugnant to me when I was in academia," he says. "Now it's no big deal, and it allows life to run smoothly. Nor would I have learned about cash-flow forecasts or measuring the time value of money. These are administrative techniques for making life easier, and learning them has actually been an extremely valuable lesson."

When Doyne and Norman first got into the prediction business, they thought it would be a brief diversion. They would peer into the world financial markets and get back to thinking about more important things: the origin of life and its future prospects; the laws of evolution, complexity, and self-organization; and the relation of human beings to the systems they make. But, instead, the process has altered them. The horizon has kept receding. The project has become increasingly complex and all-encompassing. They diverged from their former selves, and found their own lives coming under the sway of chaos, whose two fundamental laws are *sensitive dependence on initial conditions* and *rapid divergence of nearby trajectories*. Norman has grown into being a manager, a negotiator, and organizer. Doyne has mastered the art of slashing through complicated problems to arrive at workable solutions. His company mantra is "Turn the crank."

On March 8, 1996, the markets suffer a precipitous hundred-point drop. The plunge is occasioned by the *good* news that in the preceding month the U.S. economy created twice as many jobs as expected.

A crash provoked by good news is always accompanied by the word "inflation." *Inflation* is the code word used by the Federal Reserve Board when they intend to put the brakes on job growth or prop up the dollar by raising interest rates. Second-guessing the Fed is not Prediction Company's business, but by the end of the day the portfolio has cleared a tidy seven-figure profit. They break out the champagne. Even Clara, the big black dog with one doting blue eye, lets out a joyous *woof!*

A few days later SBC begins trading a handful of Prediction Company stock models on the New York Stock Exchange. By the end of the year they expect to have hundreds of stock models online. They are getting good at automated model building and execution. Release day for the stock models invites the usual glitches. The satellite data feed goes down. Doug Hahn's hard drive crashes. Norman and Karen work late into the night, and by the following day, the stock models are walking on their own.

This marks a major improvement in the company's performance. Prior release days were plagued by problems, some of them lasting for months. In one case, William noticed the S&P model wasn't behaving like it was supposed to. He discovered that they had released the wrong model. The old version had been plugged in, rather than the correct one, and run for five months before anyone tracked down the mistake. "This was a wake-up call for the researchers to quadruple-check their models before handing them off to the production team," says William of development protocols that this time worked like a charm.

The one thing without which Prediction Company cannot operate is data. This is the domain of Sonia Fliri-Hummer, the dimple-cheeked Tyrolean who began working on Aztec Street as a part-time employee and now fields weekly phone calls from headhunters on Wall Street. Sonia pulls down data from everywhere in the cosmos: Bloomberg, Bridge, ADP, Quotron, DRI, Knight-Ridder, Reuters, Dow-Jones, Globex. She can tell at a glance when someone is fudging the numbers. Other people in the company rely on a color-coded trading screen designed by Stephen. It is blue when running with clean data, orange when sick, red when dead, and fuscia when sounding an alert. The system is loaded with analytical tools and time-zone charts for every market around the world. "It's like a control panel popped in all our personal computers," says Sonia. The panel helps them fly their models out the door, and with more of them getting launched and hitting their targets, the company is starting to make some good money.

Escape Velocity

Any sufficiently advanced technology is indistinguishable
from magic.
> —ARTHUR C. CLARKE

The best way to predict the future is to invent it.
> —ALAN KAY

On June 21, 1996, Prediction Company receives a fax from Schweizerischer Bankverein saying, "Hooray! Two signed copies of contract
being shipped to you today. Best Wishes, Joe." Finally, the deal
arranged in December is signed. Norman, who had been afraid that
he might have to "unhire" some of the company's new employees,
breathes a sigh of relief. Doyne throws a party at his house.

After a long winter drought, which left the desert looking scorched
beyond repair, the rains have returned. The rusty brown nubbins of
grass in Doyne's yard have sprung back to life. The goldfish pond is
newly rebuilt and planted as a wetlands with a gurgling waterfall. The
wooden deck beside the pond offers a view into the Sangre de Cristos
in one direction and down to the Rio Grande valley in another. The
Russian olives are producing their sweet evening perfume. The lawn
is nicely shaded by Chinese elms. Doyne lights the grill and mixes a
batch of mint juleps before he sits down to watch the sunset on a perfect summer evening.

Letty arrives wearing a suit, having just driven up from Albuquerque, where she is threatening to drag the city into court. The city's waste water is the fifth largest tributary to the Rio Grande. They've released enough untreated effluent and chlorine to sterilize the river; no fish exist downstream of the city. After seven hundred warnings from the Environmental Protection Agency and other governmental bodies, Albuquerque is about to get sued by Letty, who is responsible for dead fish and other assaults on the state's natural resources.

She gives Doyne a kiss and goes off to change. Chris Langton pulls up a chair. Doyne has been keeping an eye on his friend, ever since Chris got in trouble a few weeks ago. He had been under a lot of pressure, trying to finish a long-overdue book about artificial life, while organizing the annual A-life conference being held that year in Japan.

When Chris drove down to Albuquerque to catch the plane, instead of flying to Tokyo he drove into the desert. He was found three days later, raving in a hallucinatory haze and near death from dehydration. After a spell in the hospital in Albuquerque, he is back to normal and glad to be alive, but he is thinking of changing his life, maybe dropping out of academic research and going into business. SWARM, his A-life program, is proving to be good at analyzing high-dimensional systems in which thousands of variables interact with one another; several people have suggested it could be used in business or financial simulations. Chris is curious to hear about Doyne's own experience in switching from academia to business.

"It's part of my rebelling against the need to impress people," Doyne tells him. "To be free of the bureaucracy and grant grubbing is a great relief. Starting a company is a self-effacing activity. You have to play on a team, and it's a real pleasure to help create an organization of twenty people who can function together so well. In business the only way they keep score is P&L, not reputation, peer review, or all the other ego-laden intangibles that people try to measure in academia."

As summer shades into fall, the predictors are knocking off corporate objectives like bowling pins and cranking out new models by the dozens. By the time Fiesta arrives, Zozobra reveals himself to be a zoot-suited punk in a bright white dress with red buttons. The basketball-sized pompons that make up his hair are silver and this year, for the first time, a Doom Queen, also dressed in silver, is being added as a counterpoint to the Fire Dancer. Zozobra burns in a crowning fire that throws a tremendous burst of heat into people's faces. He is consumed in seconds.

Cramped for room, Prediction Company takes over the space formerly occupied by the mystical healer known as White Buffalo. They move out the tarot cards and massage oil and start channeling their own form of cosmic energy. At the end of 1996, still needing more room, the entire company moves around the corner to the old Capitol Hotel on Montezuma Street. Built in the 1880s, the Capitol is a former railroad workers' hotel and whorehouse, which is now owned by Stuart Kauffman. Kauffman is the theoretical biologist with whom Norman and Doyne wrote two papers on self-organizing systems and the origins of life. Apart from dabbling in real estate, Kauffman is also getting into the prediction business. He has founded a company called Bios, which uses nonlinear dynamics and chaos theory to model complex systems. Their first big contract is to model the NASDAQ stock exchange.

The Capitol Hotel is a solid brick building with fourteen-foot ceilings and big bay windows looking out to Santa Fe Baldy. It has ceiling fans and Tiffany glass chandeliers. It even comes complete with a ghost named Elizabeth, a prostitute who met an untimely death by falling down the hotel's steeply pitched stairs. After the building is rewired into a high-bandwidth hive, the software engineers claim the top floor, which has the best views onto the mountains, the researchers get the bay windows on the second floor, and the ground floor is loaded with computers. Laura fills the building with flowering cacti and other plants in ceramic pots. There is even a second-story kitchen and outdoor deck overlooking the Kokoman Circus liquor store and Double-A jazz club.

Installed on Montezuma Street, Prediction Company starts the new year with a "tough slog" toward the release of their 2.6 software, which is designed to run a thousand predictive models. The new hires are settling in. The headhunters are calling, but getting nowhere. "It's a superb team," Doyne says. "We should be able to crack this problem."

The slog continues until May 1997, with people working late nights to get the new trade engine into production. On release day they flip the switch and the software runs without a glitch. Finally, they are learning how to underpromise and overperform.

May is also the month that the boa constrictor known as Swiss Bank swallows another mouse. They announce the acquisition of Dillon Read & Co. in New York for six hundred million dollars. This old blue-blood investment bank will be rolled into SBC-Warburg. Swiss Bank suffers its usual indigestion after eating this corporate meal, but the Prediction Company portfolio is handled within the bank by its own production team and traders that keep it sheltered from the chaos that periodically engulfs the rest of the bank.

It is also in May that Doyne flies to Santa Cruz to see his old teachers and friends at the University of California, where he has been invited by the physics department to give a lecture. This marks the first public airing of his paper on market forces and ecology. On his way into the science building, he is amused to find installed in the lobby a display case devoted to roulette. Featured in this "little shrine" to Eudaemonic Enterprises are a toe-operated roulette computer and shoe, equations from Doyne's research notebooks, and a signed first edition of *The Eudaemonic Pie*.

Doyne's lecture is packed. The audience is enthusiastic, except for one meddlesome professor, who manages to remind Doyne why he left academia. Doyne answers a lot of questions and finally breaks away to spend the evening with Rob Shaw. Now that his quarter-million-dollar MacArthur "genius" award is gone, Rob has had to go to work. He is commuting from Santa Cruz to Silicon Valley for a job at Interval Research Corporation, a think tank financed by Paul Allen, cofounder of Microsoft. Rob's assignment is to put "presence" into computers. He is using little mechanical thumpers and other force-feedback devices that allow you to "feel" the objects on the screen.

This is a good day job, but Rob is equally excited about the company he is founding with his brother, Chris. They have taken over the old BPOE lodge and roller-skating rink in downtown Santa Cruz, divided the space with parachutes hung from the ceiling, and filled it with Chris's surreal paintings and Rob's physics toys. Called Haptek, the company is building virtual-reality bodysuits that give you a punch in the stomach when you are fighting bad guys. The punch is generated by vibrating solenoids—the technology that Eudaemonic Enterprises used to get betting signals transmitted to the soles of their feet. Haptek has another promising contract to build a synthetic Dan Rather. This is a virtual newsreader that can report events from anywhere in the world without actually being there. It is an "avatar," or animated computer image, with graphics so sophisticated that it displays a full range of three-dimensional motion, plus the face of Dan Rather, Abraham Lincoln, or anyone else suitable for reading the evening news.

The other big event in May is a Prediction Company party to celebrate their move to the Capitol Hotel. It has taken since December to get around to unboxing books and hanging pictures on the walls. Finally, after the kitchen and deck are built and Doyne's father has installed a two-hundred-watt stereo system, it is time to celebrate.

The Cowgirl Hall of Fame delivers a big spread of food. Kokoman Circus provides the drinks. Doyne shows up wearing his "Eat the Rich" T-shirt. Weinberger is there in full gunslinger regalia. Even Zeus Pelkey comes to the party. He could never visit their old office on Aztec Street because he was too proud to have someone carry him up the stairs. But the Capitol has an old turn-of-the-century elevator. The building inspector recommended this antique be used only on special occasions, but this is one of them. Pelkey rolls himself into the creaky box and gets hoisted upstairs for a tour of the premises. He installs himself on the second-floor deck, between Michael Nesmith of the Monkees, who is now dabbling in venture capital, and a crowd from the Santa Fe Institute. Stephen Pope as DJ cranks up the new speakers, and the party rocks until long past when all the kids running around the building are supposed to be in bed.

With their first-rate researchers, advanced models, fault-free execution, and voracious data crunching, Prediction Company is beginning to get a reputation as the finest black-box operation in the business. But the more attention they get, the less black the box. One day the company is surprised to receive a letter from someone who has been keeping an eye on them. The corporate author is so impressed by Prediction Company's stock-trading performance that it wants to go into business with them.

The predictors have been funneling some of SBC's trades through one of Wall Street's electronic brokers. The broker in turn funnels *its* trades through an automated execution service called DOT, which is owned and operated by the New York Stock Exchange. With a lot of resistance from the specialists who make their living shouting orders on the trading floor, DOT, which stands for "designated-order turn-around," was opened in 1976. It allowed the Big Board's two hundred member firms to connect directly to the exchange through comput-ers. Even in this tightly controlled form, with restricted access and limited hours of operation, DOT has become the backbone of the New York Stock Exchange. It currently accounts for forty percent of its volume and eighty percent of its order flow, with execution aver-aging twenty-two seconds per trade.

The next step in the automation of finance will be the development of trading networks open to anyone with a computer. This will elimi-nate the middlemen and cut the execution time and fees down to zero. "Machines are just a lot smarter and more efficient at this kind of thing than humans," claims New York investment banker David Shaw. He and other financial entrepreneurs are rushing to build their own electronic trading networks, which Shaw predicts will someday put the New York Stock Exchange out of business. The simplest of these networks is a stock crossing system that matches buy and sell orders and executes them online. A company called Instinet does this twice a day, at the close and two hours after the close of trading on the exchange. But the ultimate crossing network will do this nonstop, in real time, twenty-four hours a day.

International banks and insurers and big investors are madly scrambling to build the infrastructure for electronic finance, but it appears that a handful of people working out of an old whorehouse in Santa Fe, New Mexico, rank among the world's experts in this tech-nology. Prediction Company's portfolio management tools, software, number crunching, data acquisition, and risk analysis—all the operational stuff they invented *ex nihilo,* or built on the fly—are exactly the technology now being coveted by the big players in finance. This results in Prediction Company and their Swiss partners receiving another intriguing proposal from a major supplier of portfolio management software. The president of the company flies out to chat with Norman about the possibility of designing a "Prediction Company button" that will allow portfolio managers to execute trades directly from their computer screens.

In the meantime, Swiss Bank is also suggesting that Prediction Company get into automated market making, and another company is trying to lure them into trading securities in London. Still other people want to buy Prediction Company's analytic tools or database or software for organizing the flood of numbers that flows around the world. Just by *consulting* on the markets they could make enough money to forget about trying to beat them. These offers are flattering, and some of them are tempting, but is there really enough spare time for the predictors to get into the extraneous business of revolutionizing global finance?

"We're entering Phase III," David Weinberger declares. "The company has proved itself. You have mature technology ready for deployment. It's time to move from proprietary trading into execution and data services. This is the moment to establish yourselves as the dominant brand and seize market share."

This is heady talk for a company whose P&L is still wobbly. The satellite data feeds into Santa Fe, and even the electricity and telephones, crash with unnerving frequency, and mañana is the watchword for the local companies who are supposed to keep these services running. Jim Nusbaum is assigned the task of figuring out how these alliances might work. He likes the idea of spinning off a separate company that would allow them to capitalize on their lead, but all these questions are left hanging for the moment.

Zozobra this year is stuffed with thirty-five bags of arrest records from the police department. As an added feature, people from the community are invited to write down their gloomy thoughts and add them to the pyre. Last year's punk Zozobra is replaced by a more traditional, black-buttoned figure. On a beautifully clear evening most of Prediction Company, including friends like Jim Crutchfield, who is celebrating a new job at the Santa Fe Institute, gather on the lawn in Old Fort Marcy Park for the yearly picnic and burning. They pass around goat skins of wine until the kettle drums begin tapping out Zozobra's death march. The lights go down. Up go the shouts, "Burn him! Burn him!" The Glooms float over the hill. The Doom Queen makes her second annual appearance. The Fire Dancer mounts the Mayan steps and torches the groaning monster. "Burn him! Burn him!" Zozobra ignites in a satisfying flame that in seconds transforms him into a sparkling cloud of fire and ash. *¡Vivan las Fiestas!*

The speculation about Prediction Company lending its technology to market making, stock trading, data consulting, and other forms of automated finance is brought to a halt at the October 1997 board meeting, held at the Eldorado Hotel. Pelkey is spending a lot of time on meditation, moving inward in his fight with pain. David Weinberger has decided to go off the board and concentrate on his own health problems. So the meeting is an intimate threesome with Zeus Pelkey telling Doyne and Norman that they should get back to beating the markets and forget about these other distractions.

"I didn't invest in a service company," he says. "I invested in a company that was supposed to capitalize on predicting market moves. You guys are big thinkers, but you're thinking about more things than you can do. You only have twenty people and so many hours in the day. My bet is on research to beat the markets. So get busy and do it."

Pelkey's final advice is that Doyne should "hand off the management and focus exclusively on research." The upshot of the meeting is that all the would-be suitors are sent away and Prediction Company is restructured. Norman becomes chief executive officer, running the company by himself, and Doyne focuses exclusively on building and running the models.

"We've been slogging through the mud for six years, and this looks like the most direct route to high ground," Norman admits.

"We have all the resources we need in this building to nuke the problem," Doyne declares. "If we don't do it, we don't deserve to exist."

During their meeting with Pelkey at the Eldorado, the stock market crashes. The Dow Jones Industrial Average plunges five hundred and fifty-four points, with the market losing seven percent of its value. This is the worst day in a decade and the twelfth worst day in the market's history. The crash occurs on October 27, 1997, almost sixty-eight years to the day since the crash that marked the onset of the Great Depression; in 1929, the market plummeted more than twelve percent in a single day. The 1997 crash follows a tumultuous autumn with massive devaluations and collapsing markets in Thailand and elsewhere in Southeast Asia, but once again the experts are mystified by the day's events. Why should the markets crash? Why today?

Prediction Company's models are supposed to have winning days whether the markets go up or down; the strategy is designed to be "market neutral." Even so, it is with trepidation that Doyne and Norman hustle back to Montezuma Street after their meeting with Pelkey. They push open the door of the old hotel, which is still marked "Lobby," to find an electrified crowd gathered around the trading station. William is directing a fierce gaze at the numbers. Norman begins pacing the room. Doyne and his dog, Clara, stare at the screens. Everyone watches the green lines break up and gap downward as the markets plummet. The predictors hold their breath, waiting for the first signs of how the orders are getting filled.

Suddenly, a big cheer goes up, and they begin clapping one another on the back as the prices tick upward and the daily P&L clocks a tidy multimillion-dollar profit. This is the portfolio's best day yet. "Way to go!" shouts Doyne. "Not bad," Norman acknowledges, with a broad grin on his face.

In December 1997, the acquisitive Marcel Ospel strikes again. This time he is merging SBC with the Union Bank of Switzerland to create a company with seven hundred billion dollars in assets and one trillion dollars under management. The world's twenty-seventh largest bank is uniting with the world's seventeenth largest bank to form a company, called UBS, which will rank among the world's largest financial institutions.

The initial announcement makes the merger of Switzerland's first and third largest banks sound like the happy marriage of cousins. But as soon as they learn who will be running the new company, the financial press begins speculating that this is a shotgun marriage. Most of the key positions, save for a few dull spots in consumer banking, will be held by Marcel Ospel and his colleagues. Gary Brinson will handle money management. Foreign exchange and interest rates go to Andy Siciliano. David Solo, the former derivatives trader, will be chief operating officer of SBC-Warburg, the investment banking division. Other major jobs in investment banking—from which the combined operation expects to make most of its money—are handed out to the old crew from O'Connor & Associates. The Chicagoans have come a long way with their triple-A credit rating. So, too, has Marcel Ospel. Ever since he foresaw the rise of derivatives and the need for SBC to buy the world's best talent for playing these new markets, he has catapulted a sleepy Basel bank into a global powerhouse. For his prescience, he is named chief executive officer of the new UBS.

"They got the name. We got the keys," quips one Swiss Bank official. He is joking about the fact that SBC, while getting its traditional symbol of three crossed keys included in the new UBS logo, is also getting the keys to UBS's vaults. In the meantime, Union Bank's demoralized employees put up a mock web site, where UBS stands for "unemployed before spring."

The next break in the story comes when the press starts speculating that UBS is slipping into Ospel's hands because the company is facing losses of seven hundred million dollars in derivatives trading. They bet that Japanese bank shares would rise in price when, in fact, they were in the midst of collapsing. Their bad judgment was compounded by plugging incorrect numbers into their computer models, which resulted in selling billions of dollars of derivatives at the wrong price. UBS derivatives traders in London were in charge of setting their own risk limits as well as managing their own accounts, said the *Economist* magazine. This is the same recipe for disaster that destroyed Barings. A common thread in these stories about banks and other institutions getting blown up by derivatives is how few people really know how to measure risk and compute the odds of winning or losing. Without this knowledge, playing the world markets is a bad bet.

Four months after Marcel Ospel's coup, John Reed at Citicorp—who helped provoke Norman and Doyne into thinking about financial markets in the first place—announces a coup of his own. Regardless of the fact that it is currently illegal, because it contravenes the Depression-era laws designed to keep insurance companies protected from the vagaries of Wall Street, Reed is merging Citicorp with Travelers Insurance to create an entity called Citigroup. With seven hundred billion dollars in assets, it will be the world's second largest bank, behind UBS. Reed assumes, and he is probably correct, that from now on Wall Street will be making its own rules.

While Swiss Bank is getting bigger, Prediction Company is getting smarter. In mid-December 1997 they release a new 2.8 trade engine, which allows for fully automated trading on the New York, NASDAQ, and other exchanges. "It's been grueling, but everyone did a fantastic job," Doyne reports. The predictors take a week's vacation between Christmas and New Year's for a much-needed break. They return to work in January to find that the portfolio has been racking up a steady succession of wins.

"I'm much happier now that I have my hand on the modeling throttle," Doyne says. "I never should have allowed myself to get suckered into management and meetings and organizing things."

Prediction Company's strategy going forward is to model all the world's stocks, at least the ones that are traded with some frequency, and then double back and rebuild the company's foreign exchange and futures models. After their initial success, they had suffered a bumpy ride, and Norman and Doyne could never shake the fear that these models, built in haste, had become overfit from peeking too many times into the future. One suspects this is the problem when a model's live trading never matches the promise of its simulations. The company's stock models, on the other hand, look better and better all the time.

"We're trying to gather data from all the markets and integrate these information streams into one system," Doyne explains. "Not every model is a home run, but if you keep hitting singles and scoring, you don't have to hit a home run every time."

"The software is tight and rolling along," he reports during a phone call. "The P&L is up. We're printing money. It feels like a breakthrough."

The process of fine-tuning the models continues. Norman is designing filter models that are good at weeding out losers. Doyne and the research group are cranking out new models for a few thousand of the most actively traded stocks in the United States. They are even thinking about building a black box for doing interbank currency trading electronically. The idea is to trade all the world's foreign currencies in a rapid-fire system that simultaneously buys and sells the same currency. The world has a dozen banks with pockets deep enough to play this game, and Warburg Dillon Read—the UBS investment bank that actually manages the Prediction Company portfolio—is one of them.

The world's major financial institutions have all joined this race to develop black-box financial systems. The bulge of credit that moves around the globe with the sun and the time value of money, whose bits and bytes are flowing with ever-increasing velocity, dictate that success goes to the swift. Moving with equal rapidity are the risks that can wipe out unsuspecting players. "No one really knows how to eliminate all the risks," says Andrew Lo, who directs the financial-engineering program at MIT. "But we do know that less sophisticated technology will lose out over time to more sophisticated technology. This is why the old-boy network is being replaced by the computer network. Call it revenge of the nerds, but everyone on Wall Street is scrambling to develop computer-driven trading programs." This is why physics and finance are converging into the new science of phynance.

Success in this endeavor can have paradoxical results. For one thing, success invites imitation, which in the world of black-box financial forecasting is a liability. Your edge is dulled as competitors pile on to your strategy, and if the strategy is too widely adopted it is no longer useful for playing the market; instead, it becomes the market.

To continue winning, one has to keep a few steps ahead of the competition. "It takes a huge amount of work to build this kind of infrastructure, particularly here in Santa Fe," Norman says, evaluating Prediction Company's progress over the years. "We went from not knowing what a stock option was to being market players. We pulled it off. We did everything we set out to accomplish, save for making ourselves filthy rich, but, heh, tomorrow is another day."

Norman and Doyne pay a visit to Jim McGill soon after he starts his new job at Morgan Stanley. They are in New York for a quarterly meeting with Swiss Bank. They find themselves with some free time and join McGill for lunch. He shows them around the company's new building on Broadway, just off Times Square, where McGill is head of engineering in an IT department with two thousand employees. They ascend to the executive dining room on the forty-first floor. Doyne gets in trouble for not having a necktie, until the rules are bent, and he is allowed to tuck into the grilled pompano. Below them stretches a commanding view of the city from Wall Street to Ellis Island. McGill regales them with stories about the fast-paced world of New York finance, where one can make tons of money, but he doesn't seem to have much time to enjoy his money. As Doyne and Norman walk out the door, they are glad it was McGill and not they who made it to the top.

In the fall of 1998 financial markets around the world begin to tumble, along with the speculators who play them. George Soros loses two billion dollars in a bad bet on Russia. David Shaw loses several hundred million dollars of the Bank of America's money. And UBS takes losses, too. It turns out to be a major investor in Long-Term Capital Management, a Greenwich, Connecticut, hedge fund that would have gone belly up without a three-billion-six-hundred-million-dollar rescue package brokered by Alan Greenspan, the chairman of the Federal Reserve. It is an anxious time, when even those possessing it refuse to vaunt their good fortune, but Prediction Company keeps quietly racking up its winning bets to close out their most successful year to date.

The latest reports from Santa Fe are encouraging. Norman and Grazia have bought their first house, an old adobe with a walled yard, just down the hill from Doyne and Letty's. Doyne has refurbished his Datsun 2000, which is kitted out with a cherry wood dashboard and leakless top, and he has built a new office in the yard behind his house. Quarterly objectives are being met. Financial reports are getting filed on time. Norman is doing an excellent job of management by walking around. Doyne and the model builders are on a roll, and the P&L is looking good. "We're reaching escape velocity," Doyne declares. He and Norman are anticipating the ride through state space, but they are also looking forward to spending more time in their gardens thinking about the evolution of complex systems—rather than living in the middle of one.

"We have a long list of things to do if we ever have enough money to do them," Norman says. He is talking about funding research projects—maybe even building an institute to work on complexity and other hard problems.

"Yes, it's a long list," says Doyne, glancing over at the monitors with their sawtooth lines ticking up and down.

Acknowledgments

Players in the world financial markets believe the numbers speak for themselves. But the numbers do not speak for themselves. They are as opaque as any other text requiring interpretation, and they rewrite themselves every second. I am indebted to the characters in this story, to whom this book is dedicated, for letting me watch them start a company and make it go. This was an exciting time, as they themselves tried to figure out what the numbers mean, and I appreciate their allowing me to share the experience.

Various traders, speculators, bankers, and other people professionally involved with money let me look over their shoulders while they played the markets. I thank them for taking the time to show me the ropes. One of these people, voicing what seemed to be the opinion of many of his colleagues, expressed his disappointment that this book was finally being published. His goal in business, he said, is "profitable obscurity." He wishes to remain anonymous, as do several of his colleagues. So the only fair thing for me to do is to leave all the traders and financiers who aided this project, as grateful as I am for their assistance, unnamed.

I bothered lots of people during the seven years spent writing this book. For illuminating those parts of the story that deal with computers and financial engineering, I am grateful to Joshua Heims, Stuart Hirshfield, Andrew Lo, David Shaw, and Joseph Traub. For explaining how fluid dynamicists take pictures of order in chaos, I thank Garry Brown and Anatol Roshko. Other people, whom I bothered daily, were the librarians and researchers whose aid was invaluable for writing a book as disparate as this. They include Glynis Asu, Sharon Britton, Joan Clair, Margaret D'Aprix, Julia Dickinson, Lynn Mayo, Catherine Miller, Kelly Rose, Kristin Strohmeyer, and Joan Wolek.

The list of editors who aided this project is long. It stretches back to my first book, where this story begins. The publishers who put that mischief in motion were Nan Talese and Jeff Seroy. The firebrands who seized the opportunity to do it again include Bill Patrick, David Sobel, and John Sterling on one side of the Atlantic Ocean, and Ravi Mirchandani, Stefan McGrath, and Michael Lynton on the other.

During the dark days when it looked as if this book would never see the light of print, it was rescued by the enthusiasm of several people at *The New Yorker*, who did a marvelous job of pushing it forward. My thanks go to Henry Finder, John Bennet, Leo Carey, Amy Tübke-Davidson, and David Remnick. I had three excellent readers for an early draft of the manuscript: Mary Mackay, Bill Pietz, and Harry Kavros. For numerous conversations about the new financial landscape, I am grateful to Kevin Kelly. For their professional support and encouragement, I thank John Brockman and Stephen Gillen. Finally, I owe a great debt to my family, who kept me sane in the midst of this extended bout of graphomania. My love and appreciation go to Roberta Krueger and Maude, Tristan, and Julian Bass-Krueger.

A NOTE ON THE TYPE

The text of this book was set in New Caledonia, a modern typeface released by Linotype-Hell in 1979. New Caledonia originates from William Anderson Dwiggens's 1938 design for linotype production— a method of setting hot metal type in lines of words or slugs. Called Caledonia, the Roman name for Scotland, Dwiggens's typeface added a touch of calligraphic style and fluidity to earlier Scotch typefaces that were designed to be set by hand, one letter at a time.

W. A. Dwiggens (1880–1956) was one of America's foremost graphic designers. He championed the revival of fine bookmaking, which began in England with the work of the nineteenth-century social reformer and printer William Morris. An accomplished puppeteer and illustrator and the designer of twelve widely used typefaces, Dwiggens is also known for his articles attacking the leading publishers of his day, whom he mocked for the poor design of modern books.